Math Challenge I-A
Counting and Probability

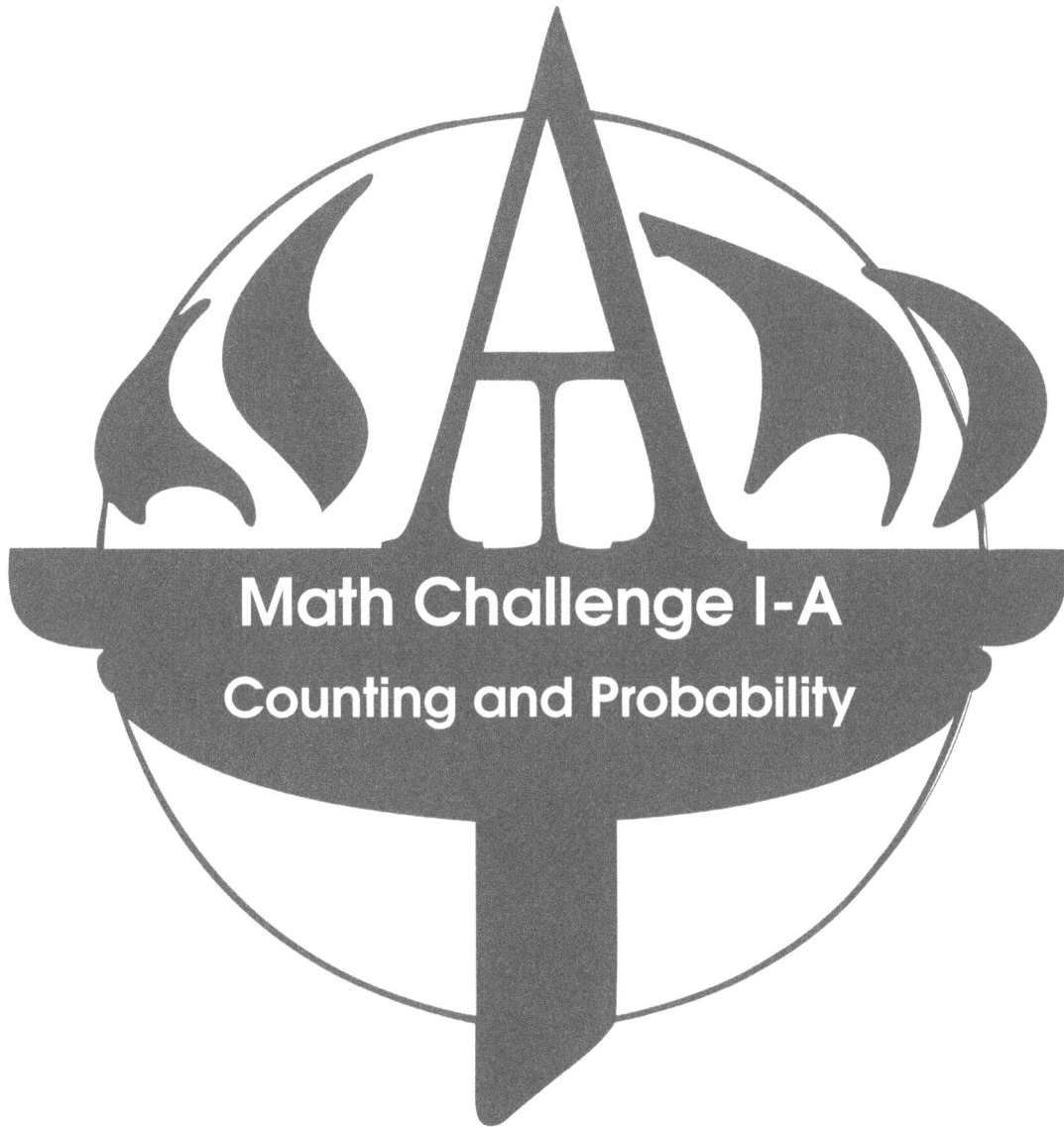

Areteem Institute

Math Challenge I-A Counting and Probability

Edited by   David Reynoso
              John Lensmire
              Kevin Wang
              Kelly Ren

PUBLISHED BY ARETEEM PRESS

ISBN: 1-944863-32-X
ISBN-13: 978-1-944863-32-6
First printing, November 2018.

Math Challenge III Combinatorics
Math Challenge I-B Number Theory
Math Challenge II-A Number Theory

Coming Soon from Areteem Press

Fun Math Problem Solving For Elementary School Vol. 2 (and Solutions Manual)
Counting & Probability for Middle School (and Solutions Manual) - From Common Core to Math Competitions
Number Theory Problem Solving for Middle School (and Solutions Manual) - From Common Core to Math Competitions
Other volumes in the **Math Challenge Curriculum Textbooks Series**

The books are available in paperback and eBook formats (including Kindle and other formats). To order the books, visit `https://areteem.org/bookstore`.

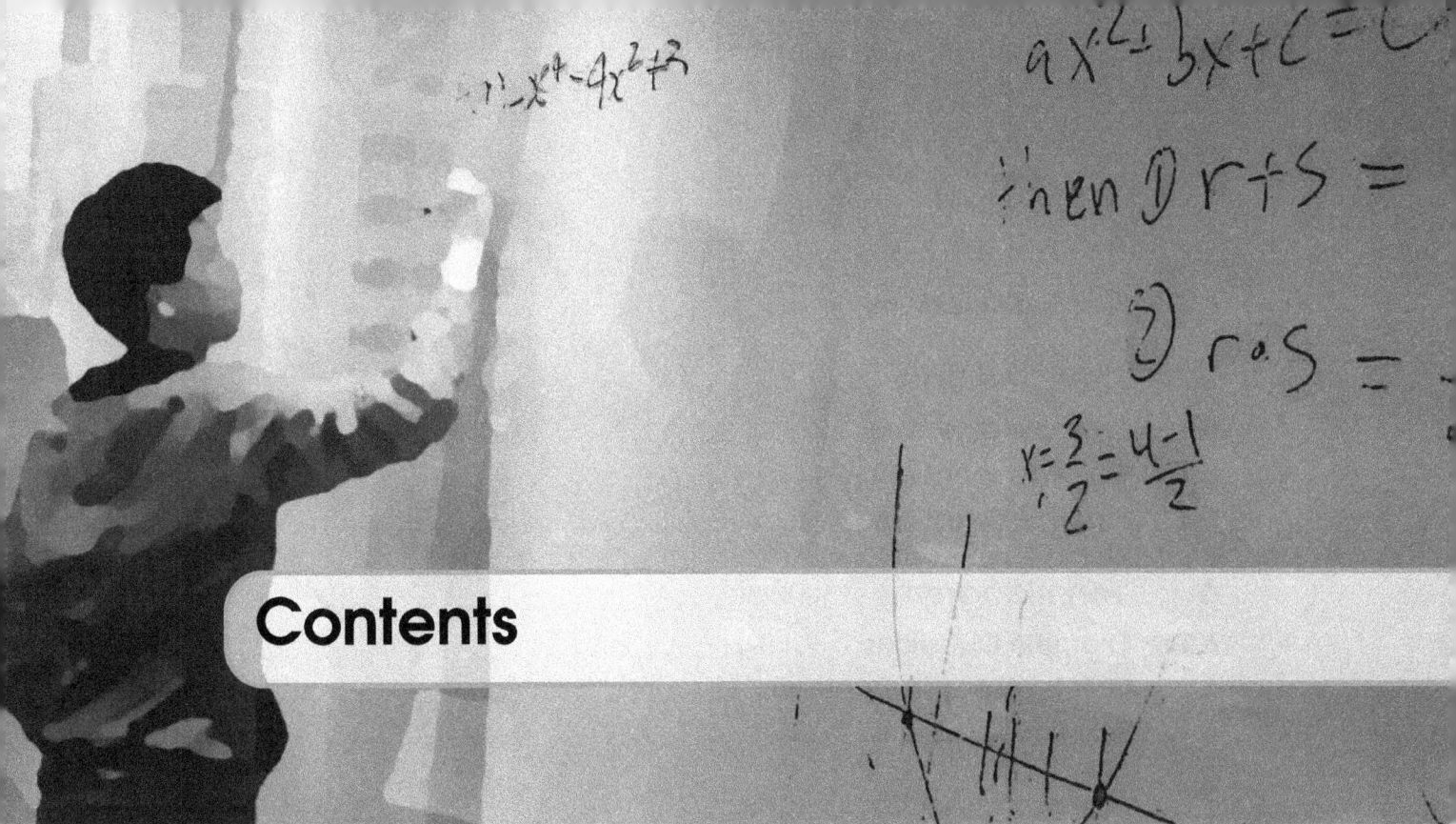

Contents

Introduction

The math challenge curriculum textbook series is designed to help students learn the fundamental mathematical concepts and practice their in-depth problem solving skills with selected exercise problems. Ideally, these textbooks are used together with Areteem Institute's corresponding courses, either taken as live classes or as self-paced classes. According to the experience levels of the students in mathematics, the following courses are offered:

- Fun Math Problem Solving for Elementary School (grades 3-5)
- Algebra Readiness (grade 5; preparing for middle school)
- Math Challenge I-A Series (grades 6-8; intro to problem solving)
- Math Challenge I-B Series (grades 6-8; intro to math contests e.g. AMC 8, ZIML Div M)
- Math Challenge I-C Series (grades 6-8; topics bridging middle and high schools)
- Math Challenge II-A Series (grades 9+ or younger students preparing for AMC 10)
- Math Challenge II-B Series (grades 9+ or younger students preparing for AMC 12)
- Math Challenge III Series (preparing for AIME, ZIML Varsity, or equivalent contests)
- Math Challenge IV Series (Math Olympiad level problem solving)

These courses are designed and developed by educational experts and industry professionals to bring real world applications into the STEM education. These programs are ideal for students who wish to win in Math Competitions (AMC, AIME, USAMO, IMO,

ARML, MathCounts, Math League, Math Olympiad, ZIML, etc.), Science Fairs (County Science Fairs, State Science Fairs, national programs like Intel Science and Engineering Fair, etc.) and Science Olympiad, or purely want to enrich their academic lives by taking more challenges and developing outstanding analytical, logical thinking and creative problem solving skills.

Math Challenge I-A is an introductory level course for 6-8 grade students who have little or no experience in in-depth problem solving nor math competitions. Students learn skills to apply the concepts they learn in school math classes into problem solving. Content includes pre-algebra, fundamental geometry, counting and probability, and basic number theory. Students develop skills in creative thinking, logical reasoning, analytical and problem solving skills. Students are exposed to beginning contests such as AMC 8, MathCounts, Math Olympiads for Elementary and Middle School (MOEMS), and Zoom International Math League (ZIML) Division M.

The course is divided into four terms:

- Summer, covering Pre-Algebra and Word Problems
- Fall, covering Geometry
- Winter, covering Counting and Probability
- Spring, covering Number Theory

The book contains course materials for Math Challenge I-A: Counting and Probability.

We recommend that students take all four terms. Each of the individual terms is self-contained and does not depend on other terms, so they do not need to be taken in order, and students can take single terms if they want to focus on specific topics.

Students can sign up for the live or self-paced course at `classes.areteem.org`.

About Areteem Institute

Areteem Institute is an educational institution that develops and provides in-depth and advanced math and science programs for K-12 (Elementary School, Middle School, and High School) students and teachers. Areteem programs are accredited supplementary programs by the Western Association of Schools and Colleges (WASC). Students may attend the Areteem Institute in one or more of the following options:

* Live and real-time face-to-face online classes with audio, video, interactive online whiteboard, and text chatting capabilities;
* Self-paced classes by watching the recordings of the live classes;
* Short video courses for trending math, science, technology, engineering, English, and social studies topics;
* Summer Intensive Camps held on prestigious university campuses and Winter Boot Camps;
* Practice with selected free daily problems and monthly ZIML competitions at ziml.areteem.org.

Areteem courses are designed and developed by educational experts and industry professionals to bring real world applications into STEM education. The programs are ideal for students who wish to build their mathematical strength in order to excel academically and eventually win in Math Competitions (AMC, AIME, USAMO, IMO, ARML, MathCounts, Math Olympiad, ZIML, and other math leagues and tournaments, etc.), Science Fairs (County Science Fairs, State Science Fairs, national programs like Intel Science and Engineering Fair, etc.) and Science Olympiads, or for students who purely want to enrich their academic lives by taking more challenging courses and developing outstanding analytical, logical, and creative problem solving skills.

Since 2004 Areteem Institute has been teaching with methodology that is highly promoted by the new Common Core State Standards: stressing the conceptual level understanding of the math concepts, problem solving techniques, and solving problems with real world applications. With the guidance from experienced and passionate professors, students are motivated to explore concepts deeper by identifying an interesting problem, researching it, analyzing it, and using a critical thinking approach to come up with multiple solutions.

Thousands of math students who have been trained at Areteem have achieved top honors and earned top awards in major national and international math competitions, including Gold Medalists in the International Math Olympiad (IMO), top winners and qualifiers at the USA Math Olympiad (USAMO/JMO) and AIME, top winners at the

Zoom International Math League (ZIML), and top winners at the MathCounts National Competition. Many Areteem Alumni have graduated from high school and gone on to enter their dream colleges such as MIT, Cal Tech, Harvard, Stanford, Yale, Princeton, U Penn, Harvey Mudd College, UC Berkeley, or UCLA. Those who have graduated from colleges are now playing important roles in their fields of endeavor.

Further information about Areteem Institute, as well as updates and errata of this book, can be found online at http://www.areteem.org.

Acknowledgments

This book contains many years of collaborative work by the staff of Areteem Institute. This book could not have existed without their efforts. Huge thanks go to the Areteem staff for their contributions!

The examples and problems in this book were either created by the Areteem staff or adapted from various sources, including other books and online resources. Especially, some good problems from previous math competitions and contests such as AMC, AIME, ARML, MATHCOUNTS, and ZIML are chosen as examples to illustrate concepts or problem-solving techniques. The original resources are credited whenever possible. However, it is not practical to list all such resources. We extend our gratitude to the original authors of all these resources.

1. The Sum Rule: Intro to Counting

Counting by Enumeration

- It is a natural tendency for us to count using our fingers: $1, 2, 3, \ldots$.
- To count how many ways something can happen, we can list out all the possibilities, and then count them up. We will call this method enumeration, a fancy word for "counting."
- It will be useful to organize the way we enumerate the outcomes to avoid mistakes. Further, enumeration can often help us find patterns we may not have noticed when we first started the problem.

The Sum Rule

- If a task can be completed using several different cases, and each case has a certain number of way to accomplish the task, then the total number of ways to complete the task can be computed by adding up the number of ways for each case.
- For example, if we wanted to count the number of sandwiches available at a fast food restaurant, we could divide the sandwiches into categories of burgers, chicken sandwiches, fish sandwiches, etc, and count each case separately.
- The sum rule is often associated with the word "or". For example, if we do "this OR that", then we would add.
- For more involved counting problems, it is good practice to break down the problem into basic parts and find the count for each part accordingly.

1.1 Example Questions

Problem 1.1 In the United States, how many state names begin with the letter "M"?

Problem 1.2 Suppose the numbers $0, 1, 2, \ldots, 9$ are represented as on a digital clock as shown below:

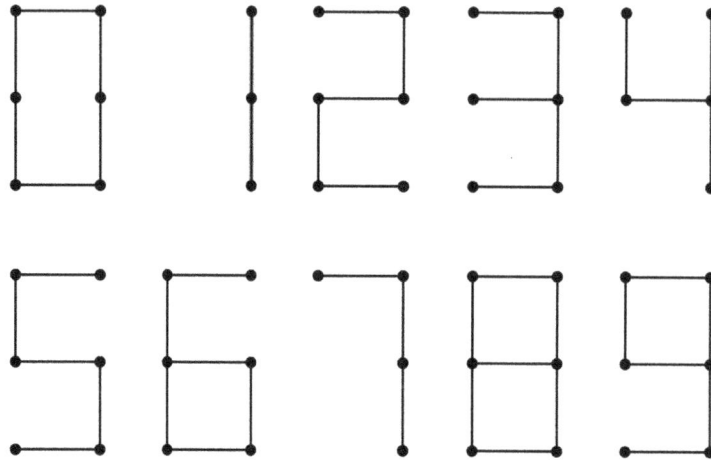

Of these digits, how many numbers have

(a) a horizontal line of symmetry?

(b) a vertical line of symmetry?

(c) both horizontal and vertical lines of symmetry?

(d) neither horizontal or vertical lines of symmetry?

Problem 1.3 How many one-digit positive integers are there?

Problem 1.4 How many 2-digit integers are there?

Problem 1.5 Suppose you flip a fair coin three times. How many possible outcomes are there?

Problem 1.6 Suppose you flip a coin a total of 3 times.

(a) Make a list of all outcomes with exactly 2 heads.

(b) Make a list of all outcomes with exactly 2 tails.

Problem 1.7 Adam and Bob played ping-pong. They agreed that whoever won 3 games first was the final winner. Assume that Bob was the final winner. In how many ways could the games have been played out?

Problem 1.8 Suppose Adam and Bob have a ping-pong rematch. They agreed to play until one person wins three games *in a row*. Suppose this time it takes 6 games to determine a winner, and the winner is Adam. In how many ways could the games have been played out?

Problem 1.9 Suppose John has 3 colored shirts (red, white, and blue) and 2 pairs of jeans (light-washed, dark-washed). Each outfit consists of one shirt and one pair of jeans. How many possible outfits can John wear?

Problem 1.10 Later that day, John decides to get some lunch. He went to his favorite Asian restaurant where customers are allowed to customize their rice plate with a choice between 2 types of rice (white, brown), 3 types of meat (chicken, beef, shrimp), and 2 choices of vegetables (broccoli, string beans). Assuming a rice plate consists of one choice of rice, meat, and vegetables, how many different rice plates can John choose from?

1.2 Quick Response Questions

Problem 1.11 In the United States, how many state names begin with the letter "C"?

Problem 1.12 How many colors are on a standard Rubik's Cube?

Problem 1.13 In a standard 52 card deck, how many face cards are colored black?

Problem 1.14 How many positive even numbers less than or equal to 30 are divisible by 3?

Problem 1.15 How many vowel characters are in this question?

Problem 1.16 How many consonant characters are in this question?

Problem 1.17 How many seasons are in a year?

Problem 1.18 How many months are in a year?

Problem 1.19 In an egg carton, eggs are arranged in 2 rows of 6 eggs. How many eggs are in 3 egg cartons?

Problem 1.20 How many letters are in the alphabet?

1.3 Practice Questions

Problem 1.21 A standard Pinochle deck consists of two copies of each card with rank 9, 10, J, Q, K, and A of all suits from a standard 52-card deck. How many cards are in a standard Pinochle deck?

Problem 1.22 In the United States, how many states begin with the letter "N"?

Problem 1.23 In the English alphabet,

A, B, C, D, E, F, G, H, I, J, K, L, M, N, O, P, Q, R, S, T, U, V, W, X, Y, Z,

how many letters have:

(a) a horizontal line of symmetry?

(b) a vertical line of symmetry?

(c) both horizontal and vertical lines of symmetry?

(d) neither horizontal or vertical lines of symmetry?

Problem 1.24 How many different ways can you flip a fair coin 5 times such that the fifth toss is a head and no two consecutive coin flips are heads?

Problem 1.25

(a) Frank needs to travel from New York to Los Angeles. He can either fly by airline, take a train, or take a bus. There are 4 airlines available, 2 train lines to choose from, and 3 bus lines to select. How many different routes does Frank have to choose from?

(b) Now suppose Frank is traveling round trip (so New York to Los Angeles and then back to New York). He still has the same options available as in (a), but he will use the same mode of transportation to and from Los Angeles (that is, if he flies to Los Angeles, he will fly back to New York, just not necessarily on the same airline). How many different round trips are possible?

Problem 1.26 The number 3 can be expressed as a sum of one or more positive integers in four ways, namely, as 3, $1+2$, $2+1$, and $1+1+1$. How many ways are there to express 4?

Problem 1.27 There are ___ e's in this sentence.

Problem 1.28 If we write out the numbers between 1 to 100, how many times would you write the digit 7?

Problem 1.29 In a mysterious room, there are two doors. Behind the first door lies 3 additional doors and behind the second door lies 4 additional doors. How many doors in total are in this house?

Problem 1.30 George has 3 cards, colored red, blue, and yellow. How many ways can you be dealt two cards, a first card and a second card?

2. The Product Rule

Review of the Sum Rule

- If a task can be completed using several different cases, and each case has a certain number of way to accomplish the task, then the total number of ways to complete the task can be computed by adding up the number of ways for each case.
- The sum rule is often associated with the word "or". For example, if we do "this OR that", then we would add.

Product Rule

- If a task can be broken down into a sequence of steps, with each step having a definite number of ways to accomplish, then the total number of ways to complete the full task is the product of the number of ways to complete each of the steps.
- For example, if we are choosing a dinner from a menu that includes 3 appetizers and 4 entrees, there are $3 \times 4 = 12$ total possibilities for a meal consisting of one appetizer and one entree.
- The product rule is often associated with the words "and" or "then". For example, if we do "this AND THEN that", then we would multiple.

Subtracting and Dividing

- Both of the sum rule and the product rule are "reversible".
- For example, suppose Tony has 30 outfits each with one shirt and one pair of pants. If Tony has 10 shirts, then he has $30 \div 10 = 3$ pairs of pants.
- The reverse of the sum rule is often referred to as "complementary counting".

2.1 Example Questions

Problem 2.1 A restaurant has 5 choices of appetizers, 3 choices for desserts, and for entree they offer 7 types of burgers and 4 types of salads. How many different meals (appetizer, entree, dessert) are available?

Problem 2.2 Suppose the Martian alphabet has 10 letters, and a Martian word is any sequence of 5 letters, allowing repeats.

(a) How many Martian words are there in total?

(b) How many Martian words are there with no repeating letters?

(c) How many Martian words are there with at least one pair of repeating letters?

Problem 2.3 Suppose that a restaurant sells 7 different burgers and 4 different salads. Two people decide to order different things, but both order a burger or both order a salad. How many different ways can this happen?

Problem 2.4 There are 40 lottery balls labeled from 1 to 40. How many ways are there to draw 5 lottery balls, in order one after another, if we replace the ball after each pick? (That is, it is possible to pick the same ball more than once.)

Problem 2.5 Consider three cities: City A, City B, and City C. There are 5 one-way roads from $A \to B$, 4 one-way roads from $B \to A$, 3 one-way roads from $B \to C$, and 2 one-way roads from $C \to A$ (and no other roads connect the cities). How many loops are there from City A back to City A (without visiting any other city twice)? Note a loop must leave City A, but does not need to visit all the other cities.

Problem 2.6 Suppose Tom from Texas is decided where to go on vacation. He can choose to go to Los Angeles, New York, or Chicago. There are 4 airlines from his town in Texas that fly to Los Angeles, 6 that fly to New York, and 8 that fly to Chicago (and the same number of airlines fly back to Texas from each city). How many total round trips are possible?

Problem 2.7 Suppose a football team has 15 members. The team plays two games, and must choose a captain and co-captain for both games. Suppose that the captain for the first game cannot be a captain or co-captain for the second game (but the first co-captain can be either captain again). How many ways can the team choose a captain and co-captain for the two games?

Problem 2.8 How many ways are there to put a white and black rook on a 8×8 chessboard so that neither can attack the other? (Rooks can only attack along rows and columns but not along the diagonals.)

Problem 2.9 Suppose you put 3 rooks (white, black and gray) on a chessboard (so none of the rooks attack each other). The white rook is in the top row of the board, the black rook is in the bottom row of the board, and the gray rook is in one of the remaining rows. How many ways can this be done?

Problem 2.10 In chess, the king can only move one space in any chosen direction. How many ways are there to put a white king and a black king on the chessboard so that they do not attack each other?

2.2 Quick Response Questions

Problem 2.11 You are allowed to choose 1 present from one of 2 large boxes that contain 5 and 15 presents respectively. If all the presents are different, how many choices do you have?

Problem 2.12 You are allowed to choose 1 present from each 2 large boxes that contain 5 and 15 presents respectively. If all the presents are different, how many choices do you have?

Problem 2.13 Alice, Bob, and Charlie run a race. If there are no ties, how many different outcomes of the race are there?

Problem 2.14 Chris and his mom each pick a number from $1, 2, 3, \ldots, 10$ (possibly the same number). How many different pairs of numbers can they pick?

Problem 2.15 Chris and his mom each pick a number from $1, 2, 3, \ldots, 10$ (possibly the same number). Suppose their numbers add up to an even number. How many different pairs of numbers can they pick?

Problem 2.16 Chris and his mom each pick a number from $1, 2, 3, 4, 5$ (possibly the same number). Suppose their numbers add up to an even number. How many different pairs of numbers can they pick?

Problem 2.17 George rolls a 6-sided die. Candace also rolls the die, but gets a roll lower than George. How many different ways can this happen?

Problem 2.18 Suppose a 8×8 checkerboard is colored red and black. How many ways are there to put a red checker and a black checker on the board so that they are in (different) squares of the same color?

Problem 2.19 Suppose a 8×8 checkerboard is colored red and black. How many ways are there to put a red checker and a black checker on the board so that the black checker is in a red square and the red checker is in a black square?

Problem 2.20 Suppose a 8×8 checkerboard is colored red and black. How many ways are there to put a red checker and a black checker on the board so that the checkers are in two squares of different colors?

2.3 Practice Questions

Problem 2.21 A restaurant has 4 appetizers, 8 entrees, and 3 desserts on their menu. They offer a dinner special, which includes 2 different entrees (a first and a second entree) and either 1 appetizer or 1 dessert for a special price. How many different dinner specials are there?

Problem 2.22 The Venusian alphabet has 5 letters, a Venusian word is any sequence of 10 letters, allowing repeats.

(a) How many Venusian words have no repeated letters?

(b) How many Venusian words are there with at least one pair of repeating letters?

Problem 2.23 Katie and Jamie go to the toy store. They narrow down their choices to 7 dolls and 5 model cars. Katie and Jamie each buy one toy, but they buy different types: if Katie buys a doll then Jamie buys a car, but if Katie buys a car then Jamie buys a doll. How many different ways can Katie and Jamie buy toys at the toy store?

Problem 2.24 George is playing Blackjack with a standard deck of 52 cards. The dealer gives George two cards: a first card face down and a second card face up. How many ways can George be dealt the two cards?

Problem 2.25 Suppose A, B, C, D are cities, and there are 4 roads from $A \to B$, 3 roads from $B \to C$, and 6 roads from $C \to D$. (Assume all road are one-way, in the direction of the arrows.) How many ways are there from A to D?

Problem 2.26 Tom from Texas has 4 pairs of shorts, 5 pairs of pants, and 10 shirts. For his vacation, he packs all his shirts and either all his shorts or all his pants. Tom is slightly afraid of flying, so he likes to plan what he will wear on his flights (both to his destination and then back) before the trip. How many different outfits can Tom plan to wear on his two flights? (Assume that Tom can wash his clothes during vacation.)

Problem 2.27 A basketball team has 8 players. After every game, the team awards the player with the most points. If the team plays 3 games, how many different ways could these awards be given?

Problem 2.28 How many ways are there to place 3 rooks (white, black, and gray) on a chessboard so that none of the rooks attack each other?

Problem 2.29 How many ways can you place 4 mutually non-attacking rooks on the chessboard, using only the outer edges of the board? Assume the rooks all look the same (but can still attack each other).

Problem 2.30 How many ways are there to put a white king and a black rook on a chessboard so that they do not attack each other?

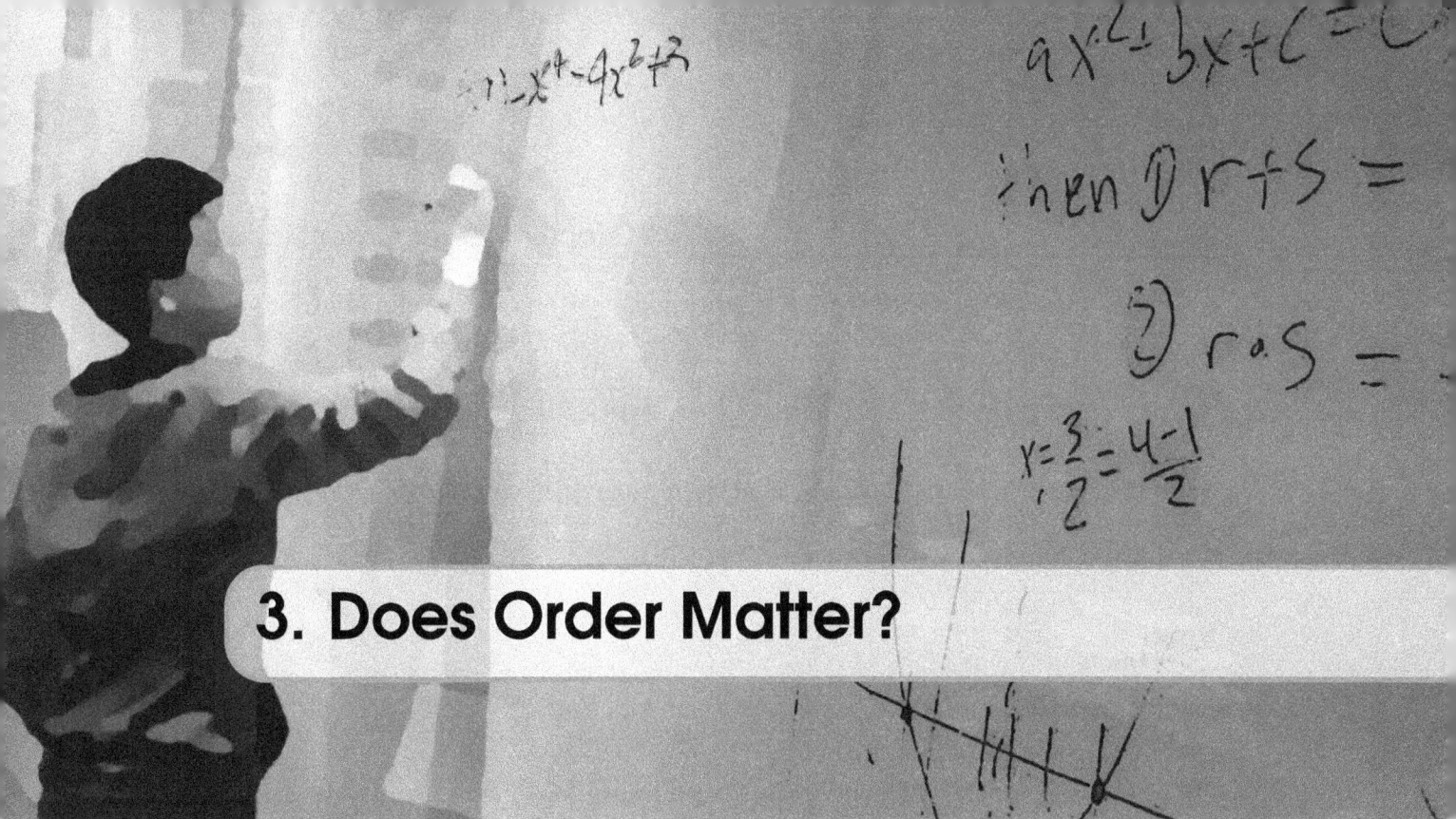

3. Does Order Matter?

Factorials

- *n*-factorial is defined as a product of all positive integers less than or equal to *n*.
- In symbols we write $n! = n(n-1)(n-2)\cdots 2 \times 1$
- For example, $4! = 4 \times 3 \times 2 \times 1 = 24$.
- We will make it convention that $0! = 1$.

Permutations

- Permutations are defined as the number of ways to order objects when order matters.
- Permutations are often written as (for ordering *k* objects out of *n* total)

$$_nP_k = \frac{n!}{(n-k)!} == n(n-1)(n-2)\cdots(n-k+1).$$

- For example, $_7P_3 = \frac{7!}{4!} = 7 \times 6 \times 5 = 210$.
- Note that $n! = {_nP_n}$.

Combinations

- Combinations are defined as the number of ways to order objects when order does not matter.

- Combinations are often written as (for choosing k objects out of n total)

$$_nC_k = \binom{n}{k} = \frac{n!}{k!(n-k)!}.$$

- Here we call $\binom{n}{k}$ a "binomial coefficient" and often read as "n choose k".

- For example $\binom{7}{3} = {}_7C_3 = \frac{7!}{3!4!} = 35.$

3.1 Example Questions

Problem 3.1 There are 40 lottery balls labeled from 1 to 40. How many ways are there to draw 5 lottery balls, in order one after another, if we do not replace the ball after each pick? (That is, it is not possible to pick the same ball more than once.)

Problem 3.2 There are 40 lottery balls labeled from 1 to 40. How many ways are there to draw 5 lottery balls all at once? (That is, it is not possible to draw the same ball twice, and the 5 balls are in no particular order.)

Problem 3.3 Suppose five people are to be seated in a row of 9 chairs. How many possible seating arrangements can be made if there must be a seat in between each person?

Problem 3.4 Suppose you have a club of 20 members. The club chooses four officers: President, Vice-President, Treasurer, and Secretary. They also choose someone to be in charge of fundraising. The four officers must all be different, but the member in charge of fundraising can be one of the officers. How many ways can we choose the four officers in the club with no restrictions?

Problem 3.5 Suppose a football team has 10 members. How many ways are there to choose 2 co-captains?

Problem 3.6 Eight friends all live in the same dorm. They receive 3 movie tickets for the weekend. They decided to draw the tickets randomly. If all the tickets are identical and a friend gets at most one ticket, how many ways can the tickets be distributed?

Problem 3.7 Suppose you flip a coin a total of 9 times.

(a) How many total outcomes are there?

(b) How many outcomes have exactly 3 heads?

(c) How many outcomes have more heads than tails? Hint: Compare this to the number of outcomes that have more tails than heads?

Problem 3.8 Suppose below is a map of a city you want to travel from A to B.

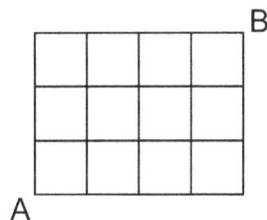

(a) If each square in the diagram is a square "block" what is the minimum number of blocks it takes to get from A to B?

(b) How many paths shortest length are there from A to B?

Problem 3.9 How many ways can you put 8 identical mutually non-attacking rooks on a 8×8 chessboard?

Problem 3.10 10 points are marked on the plane. How many different triangles can be formed using these points as vertices if

(a) no three of the points are in a straight line?

(b) 5 points are on one line, and the other 5 points are on another line parallel to the first?

3.2 Quick Response Questions

Problem 3.11 Calculate 7!

Problem 3.12 Which of the following expressions is equal to $8 \times 7 \times 6$?

$$(A) \quad 8!$$

$$(B) \quad 8! - 5!$$

$$(C) \quad \frac{8!}{5!}$$

$$(D) \quad 8! \cdot 5!$$

Problem 3.13 Calculate $\dfrac{9!}{6!}$.

Problem 3.14 Calculate $\dfrac{4!}{9!} \times \dfrac{12!}{6!}$.

Problem 3.15 Calculate $\dbinom{5}{1}$

Problem 3.16 Calculate $\dbinom{6}{2}$

Problem 3.17 Calculate $\binom{7}{3}$

Problem 3.18 Suppose 6 people run a race. How many different outcomes are there?

Problem 3.19 Suppose 6 different people compete in 4 different events. How many different ways can all the events be won?

Problem 3.20 How many ways can 4 people each choose a piece of cake from 7 total (no sharing).

3.3 Practice Questions

Problem 3.21 Suppose you have a group of 10 people. How many different photographs are there of everyone lined up if all the people look different?

Problem 3.22 Suppose 8 people run a race. If the top 4 finishers in the race can advance to the next race, how many different groups can advance?

Problem 3.23 Suppose there is a basketball tournament with 8 teams. Suppose each team plays each other once. How many games in total are played?

Problem 3.24 Suppose there are 10 friends who all live in the same dorm. They receive 4 tickets, to the movies Rocky I, II, III, and IV (so all 4 tickets are different). If each friend gets at most one ticket, how many ways can the tickets be distributed?

Problem 3.25 Suppose you have a soccer tournament consisting of 6 teams. If each team plays each other twice (each team gets one home game), how many games in total are in the tournament?

Problem 3.26 Using the standard alphabet, how many 3-letter words are there with no repeated letters?

Problem 3.27 Using the standard alphabet, how many 5-letter words are there with no two letters in a row repeated? (CATCH is allowed, ATTIC is not.)

Problem 3.28 To win the New York lottery, you must choose 6 numbers correctly from a set of 51 numbers, where the order of the numbers does not matter. How many ways are there to make your 6 choices?

Problem 3.29 To win the California lottery, you must choose 5 numbers correctly from a set of 51 numbers, along with a 6th number called the "Power" number. The 5 numbers do not need to be guessed in order, but you do need to correctly pick the Power number. How many ways are there to choose your 6 numbers (with one of the numbers being the Power number)?

Problem 3.30 Compare $\begin{pmatrix} 11 \\ 3 \end{pmatrix}$ and $\begin{pmatrix} 11 \\ 8 \end{pmatrix}$.

4. Grouping and Spacing

Review of Permutations and Combinations

- **Permutations**: Permutation means arrangement of things *in a certain order*. The number of permutations of k elements taken out of a set of n elements (without repeating) is denoted $_nP_k$:

$$_nP_k = n(n-1)(n-2)\cdots(n-k+1) = \frac{n!}{(n-k)!}.$$

- **Combinations**: Combination means selection of things where *order does not matter*. The number of combinations of k elements taken out of a set of n elements is denoted $_nC_k$ or $\binom{n}{k}$:

$$_nC_k = \binom{n}{k} = \frac{n(n-1)(n-2)\cdots(n-k+1)}{k!} = \frac{n!}{k!(n-k)!}.$$

Grouping and Spacing

- **Grouping**: If some objects must be together, think of them as a single object/group when arranging. Then arrange among those objects if necessary.
    - For example, if you are taking a photo with a couple that wants to stand together, first think of the couple as one 'person' and arrange. Then multiply by 2! to account for the ways the couple can be reordered.
- **Spacing**: If some objects must be separated, arrange the other objects first, and then place the objects in the spaces created by the others.

○ For example, if you are taking a photo with a group of two enemies that must stand apart, first arrange the other people, setting aside the two enemies. The other people create spaces for the enemies, and then place the enemies in those spaces.

4.1   Example Questions

Problem 4.1 How many rearrangements can be made of the letters in the following words?

(a) STOP

(b) RHYTHM

(c) BANANAS

Problem 4.2 There are 4 terminals in an airport with each consisting of 5 flight attendents. An airline is asked to select a group of 4 flight attendents chosen in the following way: randomly choose 2 terminals and per terminal, randomly choose 2 flight attendents. How many ways can this be done?

Problem 4.3 Suppose you have 8 people and you want to take a photo of everyone lined up.

(a) How many different photos are there in total?

(b) How many photos are there if two of the people are identical twins (who are also dressed the same)?

(c) How many photos are there if there is a couple who wants to stand together?

Problem 4.4 One student has 6 books and another student has 8.

(a) How many ways can they trade 3 books of the first student for 3 books of the second?

(b) Suppose now the first student gives 3 of their books to the second, and then the second gives 3 books to the first. Unfortunately, the second student has a very bad memory, so they maybe give the first student some of their own books back. How many ways can the "trade" now take place?

Problem 4.5 Suppose you have 5 men and 5 women at a dance class. How many ways are there to divide the 10 into 5 pairs if

(a) each pair is a male and a female?

(b) there are no restrictions in how the pairs are chosen?

Problem 4.6 James, Jim, and 6 other friends line up for a photograph. James and Jim do not stand next to each other. How many different photographs are possible?

Problem 4.7 Suppose you have 10 numbered balls and 5 numbered boxes. How many ways are there to put the balls into the boxes if:

(a) there are no restrictions?

(b) no box has more than 2 balls?

(c) no box has more than 9 balls?

Problem 4.8 How many rearrangements of the word *MACHINES* have no vowels next to each other?

Problem 4.9 Suppose a pizza place has 5 toppings available. You want to order 2 different 3-topping pizzas. Suppose repeated toppings are not allowed on a single pizza, and the order of the toppings on a pizza does not matter. If you only care which two pizzas you get, how many ways are there to make the order?

Problem 4.10 Suppose you have a group of 8 people. How many different photographs are there of everyone lined up if 2 of the people are identical twins and 3 of the people are identical triplets (the twins and triplets dress identically)?

4.2 Quick Response Questions

Problem 4.11 How many possibilities are there for the win, place, and show (first, second, and third) positions in a horse race with 8 horses if all orders of finish are possible?

Problem 4.12 How many rearrangements of the word *PEAT* are there?

Problem 4.13 How many rearrangements of the word *REPEATER* are there?

Problem 4.14 How many rearrangements of the letters in *STOP* are actual English words?

Problem 4.15 How many ways are there to pick a home team of 5 players and an away team of 5 players (no overlaps) from 20 total people?

(A) $\binom{20}{5} \cdot \binom{20}{5}$
(B) $\binom{20}{5} + \binom{20}{5}$
(C) $\binom{20}{5} \cdot \binom{15}{5}$
(D) $\binom{20}{5} + \binom{15}{5}$

Problem 4.16 How many permutations of the letters "ABCDEFGH" contain the string "ABC"?

Problem 4.17 How many permutations of the letters "ABCDEFGH" are there with the letters A, B, C all in a row?

Problem 4.18 How many ways can you choose 2 black cards and 3 red cards from a standard deck of cards?

Problem 4.19 How many ways can you arrange 2 different black cards and 3 different red cards so that the black cards are next to each other?

Problem 4.20 How many total ways can you draw 3 red cards and 2 black cards from a standard deck of cards so that the 2 black cards appear consecutively in the 5-card draw?

4.3 **Practice Questions**

Problem 4.21 How many rearrangements can be made of the letters in *MISSISSIPPI*?

Problem 4.22 In a high school class consisting of 5 freshmen, 6 sophomores, 10 juniors, and 4 seniors, a group is formed by selecting 2 people from each grade level. How many ways can this be done?

Problem 4.23 Suppose you have 4 males and 4 females. How many ways are there to line them all up so that the males are all together and the females are all together?

Problem 4.24 A California lottery has a set of 51 numbers to choose from. To win, you must choose a set of 4 numbers and a set of 2 Power numbers correctly (a number can be chosen at most once). If you do not have the pick the numbers in order (but Power numbers are different from regular numbers), how many ways are there to choose your 6 numbers?

Problem 4.25 Suppose you have 4 Klingons, 4 Vulcans, and 4 Andorians. How many ways are there to divide the 12 into 4 groups of 3 if each group has 1 Klingon, 1 Vulcan, and 1 Andorian?

Problem 4.26 How many ways are there to line up 10 people if three of the people are enemies and none of them can be next to each other?

Problem 4.27 Suppose you have 5 numbered balls and 10 numbered boxes. How many ways are there to put the balls into the boxes if no box has more than 1 ball?

Problem 4.28 How many rearrangements of *ARETEEM* are there if we require no 2 consecutive *E*'s in the rearrangements?

Problem 4.29 A pizzeria advertises itself for having 10 different toppings to choose from. There are having a special for two different 3-topping pizzas for $15. If the order of the two pizzas ordered in the special does not matter, how many ways are there to order two pizzas using the special?

Problem 4.30 Suppose you have a group of 8 people. How many different photographs are there of everyone lined up if the 8 people are 3 singles, a couple and the last 3 of the people are a family (2 parents and a child). The couple must be together and the family must stand together, with the child in between the parents?

5. Stars and Bars

Review of Arranging Letters

- If there are n As and $k-1$ Bs, there will be a total of $n+k-1$ letters. Then there are

$$\binom{n+k-1}{n} = \binom{n+k-1}{k-1}$$

ways to arrange the letters.

- Note the above works with any type of symbol. For example if you have n stars (the symbol $*$) and $k-1$ bars (the symbol $|$), there are

$$\binom{n+k-1}{n} = \binom{n+k-1}{k-1}$$

ways to arrange the n stars and $k-1$ bars.

Stars and Bars

- "Stars and Bars", also called "Balls and Urns" or "Sticks and Stones" is a very useful counting technique.
- A standard problem you can use stars and bars to solve is as follows: How many ways are there to put n identical balls in to k numbered boxes?
  - Use the star symbol ($*$) to denote the balls. Since we have n balls we will have n stars. For the boxes, use the bar symbol ($|$) to denote the dividers between boxes. Since there are k boxes we need $k-1$ bars.
  - Then note that any arrangement of the stars and bars corresponds a way to

---

place the n balls into the k boxes, so the number of ways is

$$\binom{n+k-1}{n} = \binom{n+k-1}{k-1}.$$

- This is just one example of a problem that can be solved with stars and bars, but there are many others!

5.1 Example Questions

Problem 5.1 Suppose that every time you flip a coin, you take one step forward (F) if you get heads and one step backward (B) if you get tails. You can represent the result by a string of letters like "FFBFBBF...." How many ways can you:

(a) take 10 steps?

(b) take 10 steps and wind up where you started?

Problem 5.2 Suppose you have 3 identical balls and 3 (different) boxes.

(a) How many different ways can you put the balls into the boxes? List all the outcomes using "stars and bars".

(b) Explain how to approach part (a) using cases (assuming you didn't know "stars and bars").

(c) If the first box must have at least one ball how many outcomes are there?

Problem 5.3 Suppose a cookie store sells 4 types of cookies: chocolate, peanut butter, ginger, and sugar. You want to buy a total of 8 cookies. Suppose the order you buy the cookies does not matter.

(a) How many ways are there to choose the cookies in total?

(b) How many ways are there to choose the cookies if you must buy at least one of each type?

Problem 5.4 Suppose you have 10 identical balls and 5 numbered boxes. How many ways are there to put the balls into the boxes if:

(a) there are no restrictions?

(b) each box has at least one ball?

(c) no box has more than 2 balls?

(d) no box has more than 9 balls?

Problem 5.5 Suppose that a person has 10 friends. He owns a movie theater (so has access to as many tickets as he wants) and wants to invite some of his friends to a movie Friday night. How many ways can he give his friends tickets if:

(a) he chooses 6 friends to give one ticket to?

(b) he chooses any number of friends (including none or all) to give one ticket to?

(c) he gives out 10 tickets in total, but it is possible that some friends get more than one ticket (so they can bring their family, or other friends, etc.)?

Problem 5.6 Let $a, b, c, d \geq 0$ be integers. How many solutions to $a + b + c + d = 15$ are there

(a) in total?

(b) with $a \geq 4$?

(c) with $a = 4$?

Problem 5.7 How many 4 digit numbers that do not contain the digit 0 are there that

(a) are there in total?

(b) have no repeated digits?

(c) have digits that sum up to 8?

Problem 5.8 You're playing a game of Farkel using six dice.

(a) How many different possible outcomes are there in rolling six dice?

(b) How many ways are there to get exactly three 6's?

Problem 5.9 How many integer solutions to the equation $a+b+c=15$ are there if:

(a) $a,b,c \geq 0$?

(b) $a,b,c \geq 1$?

Problem 5.10 How many solutions to $(a+b) \cdot (c+d) = 15$ and $a,b,c,d \geq 0$ are there in total?

5.2 Quick Response Questions

Problem 5.11 Now you put 1 ball in the first box, 3 balls in the third box and 2 balls in the fourth box, as shown in the picture.

Which of the following represents this using stars and bars?

(A) $|*||***|**||$
(B) $|*|*|*|*|*|$
(C) $*||***|**|$
(D) $|**|||*||*$

Problem 5.12 Suppose you roll six dice and you only care about what you rolled (not which die has which number). Which of the following could you use to represent the outcome in terms of stars and bars if you get two 1's, three 4's and a 5?

(A) $**|***||||*|||||$
(B) $**|||***|*|$
(C) $|**|||***|*||$
(D) $||*|||****|*****$

Problem 5.13 Consider your answer to the previous question. What outcome does $|*|*|**|*|*$ represent?

(A) One 1, one 2, two 3s, one 4, and one 5
(B) One 1, one 1, one 3 one 5, and one 6
(C) One 2, one 3, two 4s, one 5, and one 6
(D) None of the above

Problem 5.14 How many ways are there to put three identical balls into two different boxes?

Problem 5.15 If each box must contain at least one ball, how many ways are there to put the three identical balls into two different boxes? Represent the possible outcomes with stars and bars.

Problem 5.16 Suppose you roll two dice and you only care about what you rolled (not which number). Find the possible number of outcomes.

Problem 5.17 Alice, Bob, Charles, and Desiree are 4 students comparing the days of the week on which they were born. In total how many possibilities are there for the day of the week each is born?

Problem 5.18 Alice, Bob, Charles, and Desiree are 4 students comparing the days of the week on which they were born. How many possible outcomes are there if we don't care about who was born on a given day? That is, we only care how many of them are both on each day of the week.

Problem 5.19 Alice, Bob, Charles, and Desiree are 4 students comparing the days of the week on which they were born. How many possible outcomes are there if they were all born on different days? Assume we don't care about who was born on on a given day.

Problem 5.20 Alice, Bob, Charles, and Desiree are 4 students comparing the days of the week on which they were born. Suppose we do care about who was born on a given day. How many possible outcomes are there if they were all born on different days?

5.3 Practice Questions

Problem 5.21 How many subsets of $S = \{1,2,3,4,5,6,7,8\}$ (remember, a subset is just a collection of some members of S)

(a) have exactly 3 elements?

(b) contain only odd numbers?

Problem 5.22 Suppose you have five pennies and two match sticks.

(a) How many ways are there to put the pennies and match sticks in a row?

(b) How many ways are there to put five balls into three boxes?

Problem 5.23 Suppose a pizza place has 10 toppings available. The order of toppings on a pizza does not matter.

(a) How many different 3-topping pizzas are there if you are not allowed to have multiple toppings of the same type?

(b) How many different 3-topping pizzas are there if you are allowed to have multiple toppings of the same type?

Problem 5.24 Suppose you have 5 identical balls and 10 numbered boxes. How many ways are there to put the balls into the boxes if:

(a) there are no restrictions?

(b) no box has more than one ball?

(c) 2 boxes contain exactly 2 balls each?

Problem 5.25 8 friends go out to eat. The restaurant has 10 choices on the menu. Each friend orders one thing, but they decide to all share what they order with everyone.

(a) Suppose the friends all order different things. How many many different collections of food can they order as a group?

(b) How many different collections of food can they order as a group if multiple friends can order the same thing?

Problem 5.26 How many integer solutions to the equation $a+b+c=20$ are there if $6 \leq a,b,c \leq 8$? Hint: Does the ≤ 8 even matter?

Problem 5.27 How many 3-digit numbers are there:

(a) in total?

(b) whose digits are all nonzero and sum to 5?

Problem 5.28 You're playing a game of Yatzee using five dice. How many ways are there to get exactly three 6's?

Problem 5.29 How many integer solutions to $a+b+c=15$ are there if $a,b,c,\geq 4$ and we do not care about the order of a,b,c. That is, $4+5+6$ is the same as $5+6+4$, etc.

Problem 5.30 How many solutions to $(a+b) \cdot (c+d) = 15$ and $a,b,c,d \geq 0$ are there with $c+d=3$?

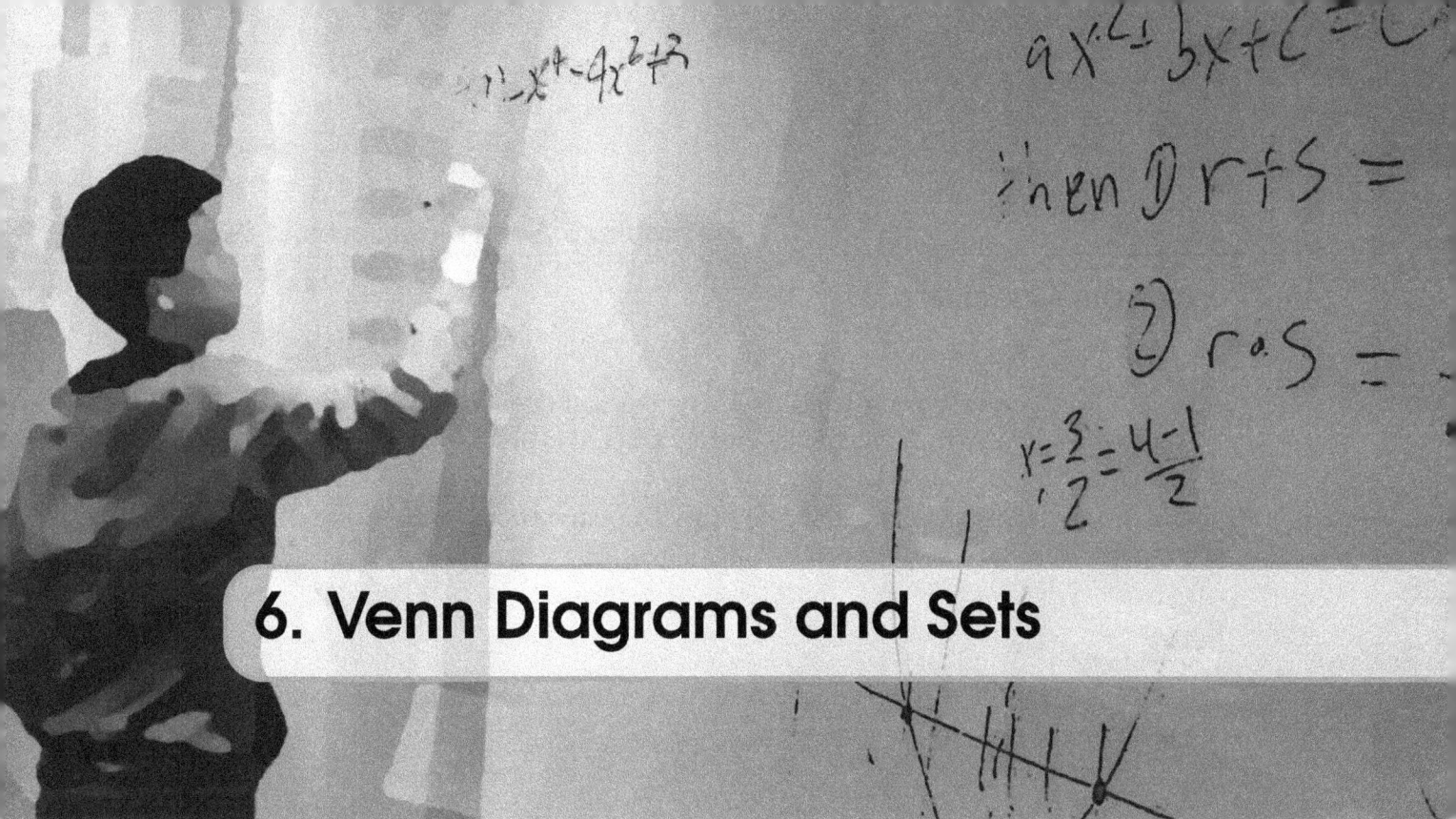

6. Venn Diagrams and Sets

Sets and Notation

- A *set* is an unordered collection of objects, without repetitions. We call the members of a set its *elements*.
- For example, $\{1,4,5\}$ is a set of 3 numbers. We have $\{1,4,5\} = \{4,1,5\} = \{1,1,4,5\}$.
- A set B is a *subset* of A, written $B \subseteq A$ if every element of B is an element of A.
- We call the set of all possible outcomes of an experiment the *sample space*, and denote it by Ω (the capital Greek letter omega).
- We will call a subset A of Ω (written $A \subseteq \Omega$) an *event*.
- In counting and probability, the complement of an event A is the collection of all elements of Ω not in A, denoted A^c.
- Recall that $A \cup B$ denotes the elements in either A or B (or both). This is called the *union* of A and B.
- Recall that $A \cap B$ denotes the elements in both A and B. This is called the *intersection* of A and B.
- The *empty set* is the set with no elements. It is denoted by $\emptyset$ or sometimes $\{\}$.
- If a set A is finite, we use the notation $n(A)$ to denote the number of elements in A.
- Note: Sets can contain more than just numbers. For example, we could have the set of all states in the US, or the set of all words starting with the letter A, etc.

6.1 Example Questions

Problem 6.1 You have five friends: two are boys and three are girls. You are blind-folded and randomly select two friends (without regard to order).

(a) Write out the list of all possible ways that this can be done. How many ways are there?

(b) What is the list of all possible outcomes that you pick two girls? How many ways are there?

Problem 6.2 Cameron wanted to play a card game with a friend using a standard 52-card deck. He offers his friend $1 if he drew an ace and $2 if he drew a spade. How many ways can Cameron's friend earn money?

Problem 6.3 Students in Areteem Institute were asked which pets (dogs or cats) do they have. In a survey of 100 students, 10 of them answered "No pets", 70 answered "a dog" and 50 answered "a cat". How many students have both a cat and dog?

Problem 6.4 How many numbers less than 100 are divisible by 5 but not 3?

Problem 6.5 Suppose you roll 2 four-sided dice. Let A be the event that the first die is a 4, and B the event that the sum of the two rolls is 6.

(a) Write out the sample space representing a list of all possible outcomes of rolling 2 four-sided dice. How many outcomes are possible?

(b) List the outcomes in event A.

(c) List the outcomes in event B.

(d) List the outcomes in event $A \cap B$.

Problem 6.6 Suppose you roll a six-sided die.

(a) What are the outcomes of the events A, B, and C, if A represents the event when you roll a 3 on a six-sided dice, B represents the event when you roll an odd number on a six-sided dice, and C represents the event when you roll a number greater than 3 on a six-sided dice?

(b) What are the events: A^c; B^c; $A^c \cap B$; $A \cup B^c$; $(A \cup B^c)^c$?

Problem 6.7 Suppose we are given a bag of 30 balls and 20 cubes. It is known that of the 50 objects in the bag, 40 of them are red and 10 of them are blue. If at least one of the objects in the bag is a blue cube, what is the minimum number of red balls in the bag?

Problem 6.8 Among 50 adults who like either coffee or tea, 30 of them like coffee and 30 of them like tea. A brand new coffee shop managed to increase the number of adults who like coffee by 20%. This decreases the number of adults that strictly love tea by a certain percentage. What is this percentage?

Problem 6.9 300 participants of the survey are asked whether they like Coke or Pepsi. According to the survey, $\frac{2}{3}$ participants like Coke and $\frac{1}{2}$ of the participants like Pepsi. Assume no one dislikes both Coke and Pepsi. How many participants like both Coke and Pepsi?

Problem 6.10 Sarah is trying to eat healthier. To do so, each day she keeps track of whether she (i) ate a salad, (ii) ate dessert.

(a) For a single day, how many different outcomes are there?

(b) For the entire week, how many outcomes are there? What might be a good way for Sarah to keep organize her data?

6.2 Quick Response Questions

Problem 6.11 Suppose you toss a coin 3 times. Let the sample space be a list of all possible results of tossing the coin 3 times. Write the elements in the sample space. What is the size of the sample space?

Problem 6.12 Suppose you toss a coin 3 times. What are all the possible ways to get 2 heads? How many possible ways are there?

Problem 6.13 Let $A = \{1,2,3,4,5\}$, $B = \{4,5,6,7,8\}$. Which of the following is $A \cap B$?

   (A) $\{1,2,3,4,5\}$
   (B) $\{4,5,6,7,8\}$
   (C) $\{1,2,3,4,5,6,7,8\}$
   (D) $\{4,5\}$

Problem 6.14 Let $A = \{1,2,3,4,5\}$, $B = \{4,5,6,7,8\}$. What is $A \cup B$?

   (A) $\{1,2,3,4,5\}$
   (B) $\{4,5,6,7,8\}$
   (C) $\{1,2,3,4,5,6,7,8\}$
   (D) $\{4,5\}$

Problem 6.15 Let $A = \{1,2,3,4,5\}$, $B = \{4,5,6,7,8\}$. What is $(A \cup B) \cap B$?

   (A) $\{1,2,3,4,5\}$
   (B) $\{4,5,6,7,8\}$
   (C) $\{1,2,3,4,5,6,7,8\}$
   (D) $\{4,5\}$

Problem 6.16 Is the set of all squares a subset of the set of all rectangles?

Problem 6.17 Is the set of all rectangles a subset of the set of all squares?

Problem 6.18 Let A be a list of all primes and B be a list of all even numbers. List all elements in $A \cap B$. How many elements are there?

Problem 6.19 Let $A = \{1, 4, 6, 7, 8, 11, 15, 19\}$ and $B = \{1, 4, 6, 7, 9, 12, 16, 19\}$. What is the size of $A \cap B$?

Problem 6.20 In a class of 45 students, 26 like to play cricket and 21 like to play football. Also, each student likes to play at least one of the two games. How many students like to play both cricket and football ?

6.3 Practice Questions

Problem 6.21 You have five friends: two are boys and three are girls. You are blindfolded and randomly select a first and a second friend (the order matters).

(a) How many outcomes are there in total? Describe the set of these outcomes.

(b) Find the number of ways that you can pick two girls.

Problem 6.22 Out of 50 enrolled students this semester, 15 students are taking Algebra Readiness at Areteem Institute and 25 students are taking MC-1 classes. If 6 students are enrolled in both Algebra Readiness and MC-1 classes, how many enrolled students are not taking Algebra Readiness and MC-1 this semester?

Problem 6.23 On Cyber Monday this year, heavily discounted products are chosen to be shipped via Amazon Prime or UPS to a small city of 2000 residents. If Amazon Prime delivered packages to 1300 residents, and UPS delivered packages to 900 residents, how many residents have packages shipped from both Amazon Prime and UPS?

Problem 6.24 How many numbers less than 100 are perfect squares but not perfect cubes? For example, $5^2 = 25$ is a perfect square and $2^3 = 8$ is a perfect cube.

Problem 6.25 Let the sample space be $\{1,2,3,4,5,6,7,8\}$ and let $A = \{1,3,5,7\}$, and $B = \{3,4,5,6\}$. Determine

(a) $A \cap B$.

(b) $A \cup B$.

Problem 6.26 Let the sample space and A and B be as in the previous question.

(a) A^c.

(b) $A^c \cap B$.

Problem 6.27 Suppose we are given a bag of 50 balls and 30 cubes. It is known that of the 80 objects in the bag, 60 of them are red and 20 of them are blue. If at least one of the objects in the bag is a blue cube, what is the minimum number of red balls in the bag?

Problem 6.28 There are objects in a bag defined by its color (red, blue) and shape (cube, ball). There are a total of 60 objects in the bag, none of which are blue cubes. 30 of the objects are red and 40 of the objects are balls. How many blue balls must be painted red in order to have a 20% increase in total number of red balls in the bag?

Problem 6.29 According to the 300 pizza lovers, 200 pizza lovers prefer that their pizza is cut in squares and 250 pizza lovers prefer that their pizza is cut in circles. This means some pizza lovers have no preference and love both square and circular pizza! What percentage of pizza lovers love both square and circular pizza?

Problem 6.30 Sarah is trying to eat healthier. To do so, each day she keeps track of whether she (i) ate a salad, (ii) ate dessert.

Sarah never has both a salad and dessert on the same day. How many different outcomes for a week (7 days) are there?

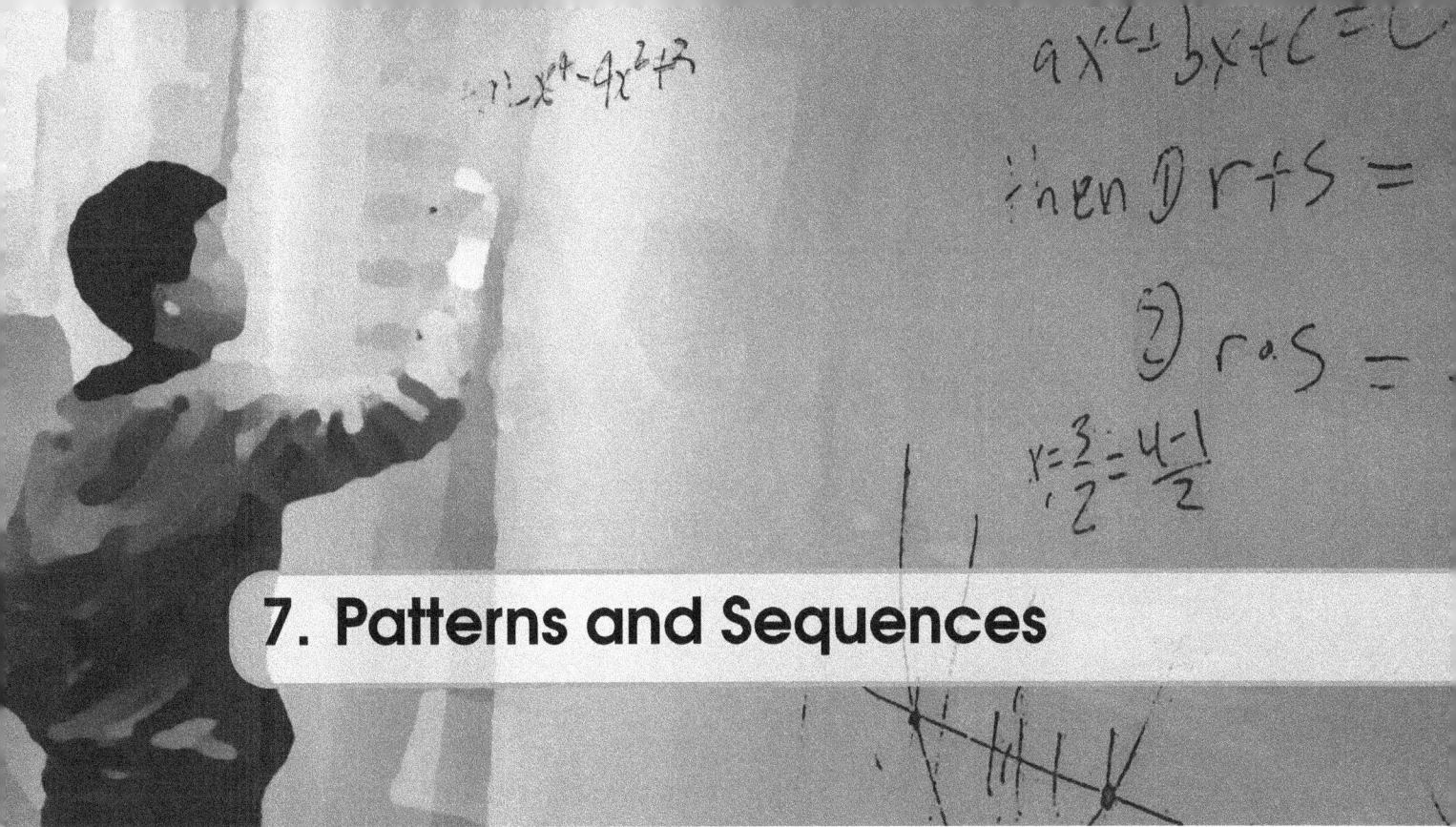

7. Patterns and Sequences

Sequences

- A *sequence* is a list of numbers (either finite or infinite) $a_0, a_1, a_2, \ldots$.
- For example:
  - $1, 2, 3, 4, 3, 2, 1$ is a finite sequence, with *length* 7.
  - $1, 2, 4, 8, 16, \ldots$ is an infinite sequence, where we have a formula: $a_n = 2^n$. Remember we start with a_0.
  - $2, 3, 5, 7, 11, \ldots$ is an infinite sequence listing all the prime numbers.
- As we've seen in the examples, not all sequences need formulas! In fact, it might be very hard to come up with formulas for some sequences.

Recursive Formulas

- Recall the sequence $3, 7, 15, 31, 63, \ldots$ from above. Each term was twice the previous term plus one. In symbols, $a_{n+1} = 2 \cdot a_n + 1$, with $a_0 = 3$.
- This type of description of a sequence is called a *recursive* formula, where the sequence is defined using the previous terms in the sequence. Note in a recursive formula you may need to specify the first term (or first few terms) of the sequence.
- For example, the sequence $0, 1, 2, 5, 12, 29, \ldots$ can be described by saying $a_0 = 0, a_1 = 1$ and $a_{n+1} = 2 \cdot a_n + a_{n-1}$. ($2 = 2 \cdot 1 + 0, 5 = 2 \cdot 2 + 1$, etc.)

Arithmetic and Geometric Sequences

- An *arithmetic* sequence has formula $a_n = a_0 + k \cdot n$ where a_0 is the starting value and k is the common difference.

- Alternatively, an arithmetic sequence has recursive formula $a_{n+1} = a_n + k$.
- For example, $2, 5, 8, 11, \ldots$ is an arithmetic sequence with formula $a_n = 2 + 3n$ and recursive formula $a_0 = 2, a_{n+1} = a_n + 3$.
- An *geometric* sequence has formula $a_n = a_0 \cdot r^n$ where a_0 is the starting value and r is the common ratio.
- Alternatively, a geometric sequence has recursive formula $a_{n+1} = r \cdot a_n$.
- For example, $3, 6, 12, 24, \ldots$ is a geometric sequence with formula $a_n = 3 \cdot 2^n$ and recursive formula $a_0 = 3, a_{n+1} = 2 \cdot a_n$.

7.1 Example Questions

Problem 7.1 For each of the following descriptions of a sequence: (i) write out the first few terms of the sequence, (ii) state whether it is an infinite or finite sequence, (iii) if it is a finite sequence, state its length.

(a) The number of days in the month, starting with January, February, etc.

(b) The sequence with where the nth term is $4n^2 + 2$.

(c) The number of ways to invite a group of 0, 1, 2, etc. friends to a party from 10 friends total.

Problem 7.2 For each of the following, (i) come up with a way to describe the sequence and (ii) write out a few more terms. Note, your description does *not* need to be a formula.

(a) $3, 1, 4, 1, 5, \ldots$.

(b) $1, 2, 9, 28, 65, \ldots$.

(c) $3, 7, 15, 31, 63, \ldots$.

Problem 7.3 Arithmetic sequences are given below, in one of three ways: (i) the first few terms of the sequence, (ii) the formula, or (iii) the recursive formula. Give the other 2 ways of describing the sequence. (That is, if the recursive formula is given, write out the first few terms and give the general formula for the sequence.)

(a) $a_0 = 5$, $a_{n+1} = 8 + a_n$.

(b) $3, 5, 7, \ldots$.

(c) $a_n = 6 - 5n$.

Problem 7.4 Geometric sequences are given below, in one of three ways: (i) the first few terms of the sequence, (ii) the formula, or (iii) the recursive formula. Give the other 2 ways of describing the sequence. (That is, if the recursive formula is given, write out the first few terms and give the general formula for the sequence.)

(a) $2, 4, 8, \ldots$.

(b) $a_0 = -3$, $a_{n+1} = -2 \cdot a_n$.

(c) $a_n = 4^n$.

Problem 7.5 More Complicated Recursive Sequences

(a) The *Fibonacci* sequence is a sequence $F_0, F_1, F_2, \ldots$ such that $F_0 = 0$, $F_1 = 1$, $F_{n+1} = F_n + F_{n-1}$. That is, the next term in the sequence is the sum of the previous two terms. Write out the first 11 terms of the sequence.

(b) Suppose a sequence starts $G_0 = 2$, $G_1 = 1$, $G_{n+1} = 2 \times G_n - G_{n-1}$. That is, multiply the previous term by 2 and subtract the term before that. Calculate the first 8 terms of the sequence.

(c) Define a sequence generated by the following: start with 12 and divide by 2 if the number is even or take 3 times the number plus 1 if the number is odd. List out the first 10 terms of the sequence.

Problem 7.6 Answer the following.

(a) In an arithmetic sequence, if the fifth term of the sequence is 5 and the tenth term of the sequence is 15, what is the first term?

(b) In a geometric sequence, if the fifth term of the sequence is 4 and the seventh term of the sequence is 9, what is the tenth term?

Problem 7.7 Calculate the Following Sums

(a) What is the sum of the first 100 positive integers?

(b) Suppose you have 30 terms of the arithmetic sequence:

$$3, 8, 13, 18, \ldots, 148.$$

What is the sum of these 30 terms?

Problem 7.8 Let a_n be the number of $n+1$ digit numbers made up of $0, 2, 4, 6, 8$.

(a) Write out a formula for a_n.

(b) Write out a recursive formula for a_n.

Problem 7.9 Let a_n (for $n \geq 1$) denote the number of ways to write n as the sum of 1's and 2's. For example, $3 = 1 + 1 + 1 = 2 + 1 = 1 + 2$ so $a_3 = 3$.

(a) Write out the first few terms of the sequence and guess a recursive formula for the sequence. (You do not need to prove your guess.)

(b) Use your guess in (a) to calculate a_{10}.

Problem 7.10 Suppose you have n friends. You want to invite some of them out to dinner. You will invite at least one friend, but not all of the friends.

(a) How many different ways can you invite a group of friends for $n = 1, 2, 3, 4$.

(b) Find a general formula with n friends.

(c) Let the answers above form a sequence $a_1, a_2, \ldots$. Find a recursive formula for a_n.

7.2 Quick Response Questions

Problem 7.11 Write the next term in the sequence: $0, 1, 8, 27, 64, \ldots$.

Problem 7.12 In an arithmetic sequence, if the first term of the arithmetic sequence is 2 and the common difference of the sequence is 3. Write out the first 5 terms of the sequence. What is the 5th term?

Problem 7.13 Suppose a sequence has recursive formula $a_0 = 2, a_{n+1} = 2a_n + n$. What is a_4?

Problem 7.14 Which of the following is a recursive definition for the sequence $a_n = 50 - 5n$?

(A) $a_0 = 45, a_{n+1} = a_n - 5$
(B) $a_0 = 45, a_{n+1} = 5 - a_n$
(C) $a_0 = 50, a_{n+1} = a_n - 5$
(D) $a_0 = 50, a_{n+1} = -5 \cdot a_n$

Problem 7.15 Find a recursive definition for the sequence $a_n = -5 \cdot (-2)^n$.

(A) $a_0 = 5, a_{n+1} = a_n - 2$
(B) $a_0 = -5, a_{n+1} = a_n - 2$
(C) $a_0 = 5, a_{n+1} = -2 \cdot a_n$
(D) $a_0 = -5, a_{n+1} = -2 \cdot a_n$

Problem 7.16 Suppose a sequence has recursive definition $a_0 = -100, a_{n+1} = a_n + 3$. Find a_{100}.

Problem 7.17 Suppose a sequence has recursive definition $a_0 = \dfrac{1}{81}, a_{n+1} = 3a_n$. Find a_8.

Problem 7.18 Suppose a sequence has formula $a_n = 2^n + n$. Find $a_6 - a_4$.

Problem 7.19 The Lucas Sequence is defined similar to the Fibonacci Sequence except the sequence begins with the first term 2 and second term 1. Therefore the Lucas Sequence starts $2, 1, 3, 4, 7, 11$. What is the first number in the Lucas Sequence that is larger than 50?

Problem 7.20 What is the sum of the first 50 odd integers?

7.3 Practice Questions

Problem 7.21 Starting with an equilateral triangle (a regular $3-$gon), write the 4 terms representing the number of diagonals in a regular $n-$gon for $n = 3, 4, 5, 6$.

Problem 7.22 Given the sequence below:

$$60, 90, 108, 120, \ldots$$

write the next 4 terms of the sequence and explain your answer. Fractions are okay.

Problem 7.23 Describe a pattern to determine the numbers in the following sequence:

$$-2, 10, 22, 34, \ldots$$

Problem 7.24 Describe a pattern to determine the numbers in the following sequence:

$$8, 12, 18, 27, 40.5, \ldots$$

Problem 7.25 Recall the Fibonacci sequence from earlier. For $n \geq 1$, let S_n be the sum of the first n terms of the Fibonacci sequence. ($S_n = F_1 + \cdots + F_n$ or $S_1 = F_1$, $S_2 = F_1 + F_2$, etc.)

(a) Write out S_n for $n = 1, 2, 3, 4, 5$.

(b) Compare your answer in (a) with the original Fibonacci sequence.

Problem 7.26 Answer the following.

(a) In an arithmetic sequence, if the fifth term of the sequence is 7 and the tenth term of the sequence is 42, what is the 100th term?

(b) In a geometric sequence, if the fifth term of the sequence is 8 and the eighth term of the sequence is 27, what is the tenth term?

Problem 7.27 The first term of the arithmetic sequence is 11 and the common difference of the sequence is 5. What is the sum of the first 100 terms in this arithmetic sequence?

Problem 7.28 Bob starts the year with 100 dollars in his savings account. Each day (including the first) Bob deposits 12 more dollars into his account. Let a_n be the amount of money Bob saves after n days (starting with $n = 0$).

(a) Write out a recursive formula for a_n.

(b) Write out a formula for a_n.

Problem 7.29 In how many ways can you express 10 as the sum of positive odd integers? For example, $10 = 3 + 3 + 3 + 1 = 3 + 7 = 7 + 3$, etc. Hint: Start with smaller numbers and find a pattern!

Problem 7.30 Start with a point. From this point draw two line segments, ending in 2 new points. From each of these new points, draw two line segments, ending in more new points. Keep repeating the process. Let a_n be the number of new points after n steps ($a_0 = 1, a_1 = 2$, etc.). Note: Such a drawing is often called a *binary* tree.

(a) What is a_3? Confirm your answer by drawing a picture.

(b) Find a general formula for a_n.

8. What is the Chance?

Set Review

- A *set* is an unordered collection of objects, without repetitions. We call the members of a set its *elements*.
- A set B is a *subset* of A, written $B \subseteq A$ if every element of B is an element of A.
- We call the set of all possible outcomes of an experiment the *sample space*, and denote it by Ω (the capital Greek letter omega). We call a subset A of Ω (written $A \subseteq \Omega$) an *event*.
- In counting and probability, the complement of an event A is the collection of all elements of Ω not in A, denoted A^c.
- Recall that $A \cup B$ denotes the elements in either A or B (or both). This is called the *union* of A and B.
- Recall that $A \cap B$ denotes the elements in both A and B. This is called the *intersection* of A and B.
- The *empty set* is the set with no elements. It is denoted by $\emptyset$ or sometimes $\{\}$.
- If a set A is finite, we use the notation $n(A)$ to denote the number of elements in A.

Probability (Classical Model)

- Suppose Ω is a finite sample space and every outcome in Ω is equally likely. If $A \subseteq \Omega$, then

$$\text{the probability of } A = P(A) = \frac{\text{want}}{\text{total}} = \frac{n(A)}{n(\Omega)}.$$

- For example, if we flip a fair coin twice, then $\{HH, HT, TH, TT\}$ is a sam-

---

ple space where every outcome is equally likely. However, the sample space {two heads, one head and one tail, two tails} is a sample space but each outcome is *not* equally likely.

Geometric Probability

- Remember that our formal formula for probability only works with a finite sample space.
- Infinite sample spaces are often difficult to work with, but one fairly easy case is when Ω is a geometric shape.
- Suppose that Ω is a geometric shape (line, square, circle, etc.) and $A \subseteq \Omega$. Then

$$P(A) = \frac{\text{want}}{\text{total}} = \frac{\text{want (in terms of length)}}{\text{total (in terms of length)}} \text{ OR } \frac{\text{want (in terms of area)}}{\text{total (in terms of area)}}.$$

- For problems involving geometric probability, it is often useful to draw a number line or a diagram.

8.1 Example Questions

Problem 8.1 Suppose you roll 2 four-sided dice. Let A be the event that the first die is a 4, and B the event that the sum of the two rolls is 6.

(a) Write out a finite sample space Ω so that every outcome is equally likely. What is $n(\Omega)$?

(b) Calculate $P(A)$.

(c) Calculate $P(B)$.

(d) Calculate $P(A \cap B)$.

Problem 8.2 Suppose you flip a fair coin 6 times.

(a) Describe the sample space Ω. Find $n(\Omega)$.

(b) Find the probability of exactly 5 heads.

(c) Find the probability of at least one tails.

(d) Find the probability of no two heads in a row and no two tails in a row.

Problem 8.3 Suppose you randomly pick a point on the number line between 0 and 4.

(a) What is the probability the number is greater than 2?

(b) What is the probability the number is greater than or equal to 2?

(c) What is the probability the number is equal to 2?

Problem 8.4 Suppose Bill has a dart board with radius 4 feet. Whenever Bill throws a dart, it randomly lands somewhere on the board.

(a) What is the probability that Bill's dart lands within 2 feet of the center of the board?

(b) What is the probability Bill's dart lands between 2 and 3 feet from the center?

Problem 8.5 A dealer starts with only the 4 aces (one of each suit) from a deck of cards. They deal you 2 of the cards. Let A be the event that you get one heart and one diamond. Let B be the event that you get a spade.

(a) Assume the cards are dealt to you in order (that is, a first card and a second card). Find $P(A)$ and $P(B)$. Hint: You may want to write out a sample space.

(b) Assume the cards are not dealt in order (that is, you are dealt two cards at once). Find $P(A)$ and $P(B)$. Hint: You may want to write out a sample space.

(c) Compare your answers in parts (a) and (b). Can you explain the outcome?

Problem 8.6 Suppose $\Omega = \{1,2,3,4,5,6\}$, and $P(1) = P(3) = .2, P(2) = P(4) = .1, P(5) = .15$.

(a) Find $P(6)$.

(b) Find the probability you get a prime number?

(c) Let $A = \{1,3,5\}, B = \{2,3\}$. Verify that $P(A \cup B) = P(A) + P(B) - P(A \cap B)$. (That is calculate each term separately and check the formula.)

Problem 8.7 Suppose you have 5 red and 8 green balls. You pick 5 without replacing the balls. For each of the events below, find the probability. Note: It is best to think of all 5 balls being picked at once.

(a) What is the probability you get 3 green and 2 red balls?

(b) What is the probability that all the balls you pick are the same color?

Problem 8.8 A, B are events. Answer the following. Drawing a Venn Diagram may help!

(a) Suppose $P(A) = .6, P(B) = .7$. What are the maximum and minimum possible values of $P(A \cap B)$?

(b) Suppose $P(A) = .5, P(B) = .6, P(A^c \cap B^c) = .2$. Find $P(A \cap B^c)$ and $P(B \cap A^c)$.

Problem 8.9 Suppose Jack and Jill decide to meet up for dinner at 5:30 PM. Unfortunately, with traffic, they each arrive (separately) sometime at random between 5 and 6 PM. Hint: think of graphing when they arrive on a square. What is the probability they both arrive on time? (Be careful what this means!)

Problem 8.10 Suppose you have two sticks, one of length 1 and the other of length 2. You now get a third stick whose length is randomly chosen between 0 and 5. What is the probability you can make a triangle with the three sticks?

8.2 Quick Response Questions

Problem 8.11 Suppose you randomly pick an integer between 1 and 20 (inclusive). The probably that your number is a multiple of 3 is $\dfrac{P}{Q}$ in lowest terms. What is $Q - P$?

Problem 8.12 Suppose you randomly pick an integer between 1 and 20 (inclusive). The probability that your number is a multiple of 3 or 5 is $\dfrac{P}{Q}$ in lowest terms. What is $Q - P$?

Problem 8.13 Suppose you randomly pick one card from a standard deck of 52 cards. The probability the card you pick is a face card is $\dfrac{P}{Q}$ in lowest terms. (Recall cards with rank J, Q, or K are called face cards.) What is $Q - P$?

Problem 8.14 A jar has 1 red, 2 green, and 3 yellow balls. Alice and Bob each pick one ball at random (at the same time so they cannot both get the same ball). The probability they both get yellow balls is $\dfrac{P}{Q}$ in lowest terms. What is $Q - P$?

Problem 8.15 Suppose Bill throws a dart at a checkerboard, and it lands randomly somewhere on the board. The probability of it landing in a black square is $\dfrac{P}{Q}$ in lowest terms. What is $Q - P$?

Problem 8.16 Suppose you have a stick of length 10 and break it once (to get 2 pieces). The probability both have length at least 3 is $\dfrac{Q}{P}$ in lowest terms. What is $Q - P$?

Problem 8.17 Suppose you have a rectangular dart board with dimensions 3 inches and 4 inches. If you have to hit a circle with diameter 2 in the center of dart board to win a round, the probability of winning a round is $\dfrac{P \times \pi}{Q}$ in lowest terms. What is $Q - P$?

Problem 8.18 Suppose $\Omega = \{1,2,3,4\}$ and $P(1) = .1, P(3) = .3$ and $P(2) = P(4)$. Find $P(4)$.

Problem 8.19 Suppose you have an unfair 6-sided die with a 50% chance of rolling a one. If the other 5 sides are all equally likely to occur, the probability that you get and odd number when you roll the die once is $\dfrac{P}{Q}$ in lowest terms. What is $Q - P$?

Problem 8.20 Suppose $P(A) = .6$, $P(B) = .5$ and $P(A^c \cap B^c) = .3$. Find $P(A \cap B)$.

8.3 Practice Questions

Problem 8.21 The slots of a roulette wheel are numbered 1-36, 0 and 00. The numbers 0 and 00 are green. For numbers in the ranges 1-10 and 19-28, the odd numbers are red and even numbers are black. For numbers in the ranges 11-18 and 29-36, the odd numbers are black and even numbers are red. Let A be the event that you get an odd number and let B be the event you get a black number on a play. Find $P(A \cap B)$.

Problem 8.22 Flip a fair coin 6 times. What is the probability of getting an odd number of heads?

Problem 8.23 Suppose you spin a needle on a circular dial (so the dial is $360°$ in total). (The needle does not necessarily stop exactly on an integer number of degrees.)

(a) What is the probability that the needle ends up between $45°$ and $135°$?

(b) What is the probability that the needle ends up at an integer multiple of $45°$?

Problem 8.24 Suppose you randomly shoot a paint ball at an 8 foot by 10 foot wall with a 3 foot by 3 foot square drawn on it. Assume that all of the paint lands on the wall and the the splatter mark is circular. If the splatter radius of the paint ball is 1 foot, what is the probability that the splatter mark is completely inside the square?

Problem 8.25 Suppose you have two sodas (each chosen from Coke, Sprite, Fanta). Let A be the event you have two of the same soda.

(a) Assume the sodas are bought in order (that is, a first soda and a second soda). Find $P(A)$. (It will be helpful to write out a sample space!)

(b) Assume the sodas are not bought in order (that is, you buy them both at once). Find $P(A)$. (It will be helpful to write out a sample space!)

(c) Compare your answers in parts (a) and (b). Can you explain the outcome?

Problem 8.26 Suppose $\Omega = \{1,2,3\}$ and $P(2)$ is twice that of $P(1)$ and $P(3)$ is three times that of $P(1)$. Find $P(1)$.

Problem 8.27 Suppose you draw 3 balls (without replacement) from a bag containing 5 red balls, 5 green balls, and 5 blue balls.

(a) What is the probability of drawing one ball of each color?

(b) What is the probability of drawing 3 balls of the same color?

Problem 8.28 Suppose A, B are events. If $P(A^c) = .3$, $P(B) = .5$, and $P(A^c \cup B) = .6$. Fill in a Venn diagram with each region labelled with its probability. That is, find $P(A \cap B^c), P(A \cap B), P(A^c \cap B), P(A^c \cap B^c)$.

Problem 8.29 Suppose Jeff, a third friend, is also coming to dinner. What is the probability that all three arrive on time?

Problem 8.30 Suppose you have two sticks, one of length 1 and the other of length 2. You randomly break the stick of length 2 into two pieces. What is the probability you can use the resulting 3 sticks to form a triangle?

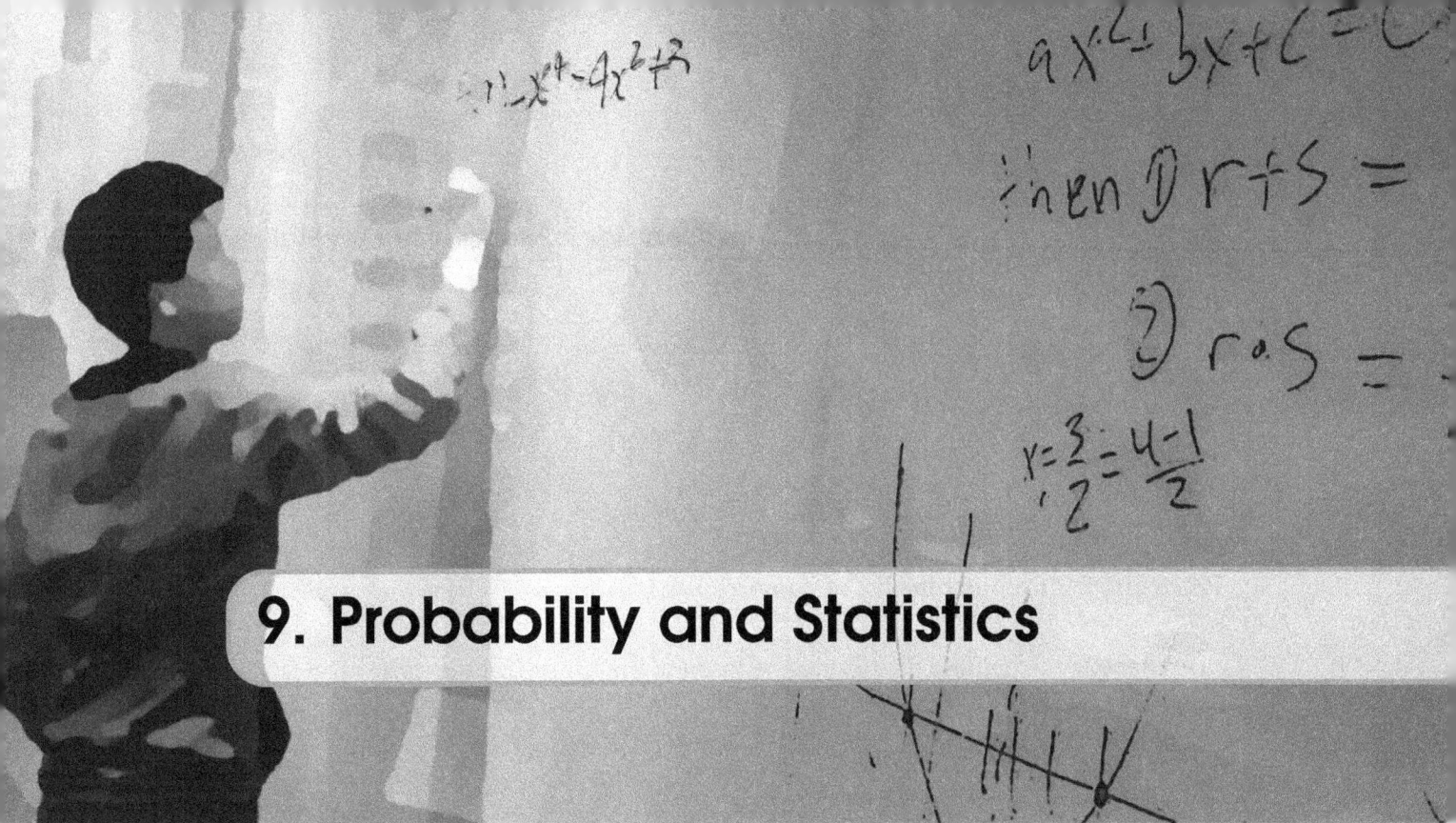

9. Probability and Statistics

Basic Statistics

- Given a list of numbers, there are different ways of calculating the "center" of the numbers.
- The *mean* of a list of numbers is the average of all the numbers. That is, you add up all the numbers and divide by how many numbers there are.
- The *median* of a list of numbers is the middle term when the list is arranged smallest to largest. If the list has an even number of terms, then the median is the average of the middle two terms.
- The *mode* of a list of numbers is the number that appears most often in the list.
- Note: Every list of numbers has a unique mean and median. However, it is possible for there to be multiple modes.

9.1 Example Questions

Problem 9.1 Given the list of numbers $1, 4, 6, 4, 3, 8, 2$ find the

(a) mean.

(b) median.

(c) mode.

Problem 9.2 Consider the list of numbers $1, 4, 6, 4, 3, 8, 2$.

(a) What (single) number do you need to insert to the list so that the mean changes to 5?

(b) Suppose instead that a new number is inserted. If the new number is an integer, what are all the possibilities for the median?

Problem 9.3 Suppose you have a list of 5 integers each between 1 and 10 (inclusive).

(a) How many different lists are possible? Answer this with and without order.

(b) How many different means are possible?

(c) Suppose the list is picked in order (first, second, etc.) one element at a time at random. What is the probability that the mean of the list is 2.

Problem 9.4 The SAT is a renowned high school standardized test administered for college admissions purposes. The scores are in multiples of 10 (maximum 800 per section) and the total score is determined by adding the critical reading subscore and the math subscore. Of the 50 students that scored above 600 on either the critical reading or math section of the SAT, 35 students scored above 600 on the math section and 25 students scored above 600 on the critical reading section. What is the maximum average total score of all 50 students?

Problem 9.5 Given below is a line graph indicating the price of a stock share in the market:

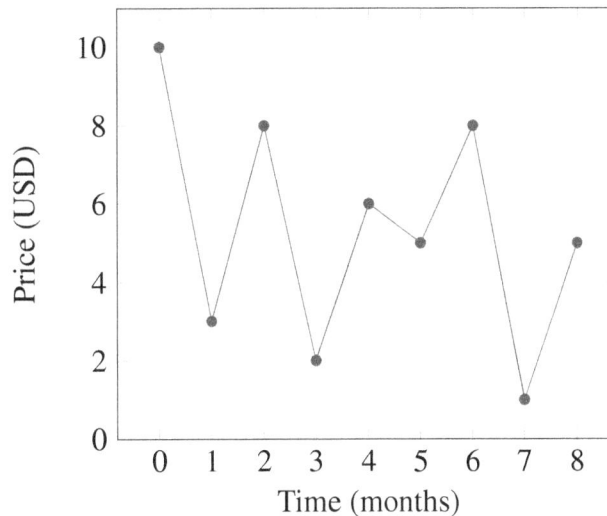

What is the average change of price of the stock every month?

Problem 9.6 Given below is a bar graph representing the ages of all students in a class.

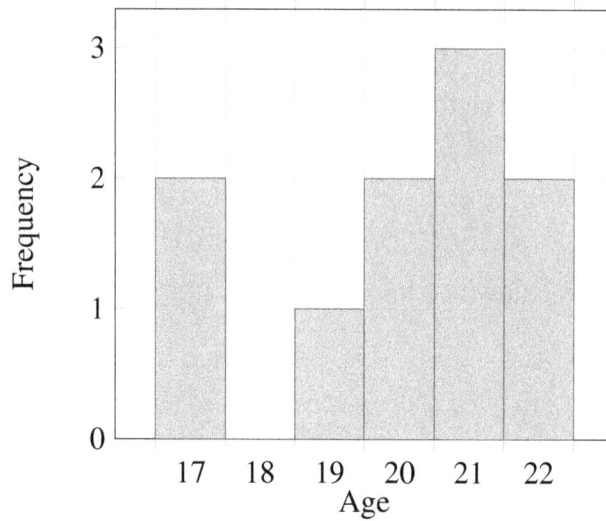

Find the median age of the students in the class.

Problem 9.7 Suppose you have the list of integers $0, 0, 2, 2, 2, 6, 8, 8, 9, 10$. You randomly pick one element in the list.

(a) What is the mode? What is the probability of picking the mode?

(b) What is the median of the list? What is the probability of picking the median?

(c) What is the probability of picking a number larger than the median?

Problem 9.8 Pretend that everyone likes all the Harry Potter books equally, so every person has an equal chance of picking any of the books as their favorite. Suppose you poll 5 people.

(a) What is the probability that the list of favorite books is $5, 2, 7, 1, 6$?

(b) What is the probability that the median of the favorite books (of the 5 people polled) is the last book? Hint: This means 3, 4, or all 5 chose the last book as their favorite.

Problem 9.9 Provided below are the heights (in inches) of 5 randomly selected adult males in the U.S.:
$$70.2, 73.9, 76.9, 69.4, 72.2$$

Provided below are the heights (in inches) of 5 randomly selected adult females in the U.S.:
$$70.6, 73.0, 65.0, 68.0, 61.9$$

The following questions regarding the heights of U.S. adult males and females are based on the data collected above.

(a) Find the mean height of U.S. adult males.

(b) Find the mean height of U.S. adult females.

(c) Find the median height of U.S. adult males.

(d) Find the median height of U.S. adult females.

(e) Is it true that it is more likely for a randomly selected U.S. adult male to be taller than a randomly selected U.S. adult female? Why?

Problem 9.10 Suppose the following list represents the grades of a test in a math class:

$$60, 64, 78, 80, 90, 92, 94, 95, 97, 99$$

(a) Find the median score of the class.

(b) Find the mean score of the class.

(c) Suppose another 5 students in the class score 90 on the quiz. How does this effect the mean and median?

(d) Do you think the median or mean does a better job at describing the given data?

9.2 Quick Response Questions

Problem 9.11 What is the mean of the following numbers?

$$10, 39, 71, 35, 76, 38, 25$$

Problem 9.12 Determine the mean of the following set of numbers:

$$40, 61, 95, 79, 9, 50, 80, 63, 109, 42$$

Problem 9.13 What is the median of the following numbers?

$$10, 39, 71, 42, 39, 76, 38, 25$$

Problem 9.14 What is the mode of the following numbers?

$$12, 11, 14, 10, 8, 13, 11, 9$$

Problem 9.15 Find the mean of the list $1, 2, 3, \ldots, 11$. Round your answer to the nearest hundredth if necessary.

Problem 9.16 Find the mean of the numbers in the list $\dfrac{11}{11}, \dfrac{11}{10}, \dfrac{11}{9}, \ldots, \dfrac{11}{1}$. Round your answer to the nearest hundredth if necessary.

Problem 9.17 Consider two lists with 11 elements each:

$$A : 1, 2, 3, 4, \ldots, 11 \text{ and } B : \frac{11}{1}, \frac{11}{2}, \frac{11}{3}, \frac{11}{4}, \ldots, \frac{11}{11}.$$

Which of the following is true the mean and medians of the two sets?

  (A)   The median of A is larger than the mean of A
  (B)   The mean of A is larger than the median of A
  (C)   The median of B is larger than the mean of B
  (D)   The mean of B is larger than the median of B

Problem 9.18 Given below is a list of the number of innings pitched by a player in the Major Baseball League:

$$5, 7, 2, 6, 8, 2, 4, 9$$

How many innings does this player have to pitch in the next game so that the average number of innings pitched in total is 6?

Problem 9.19 Given below is a line graph indicating the amount of water in a leaky bucket:

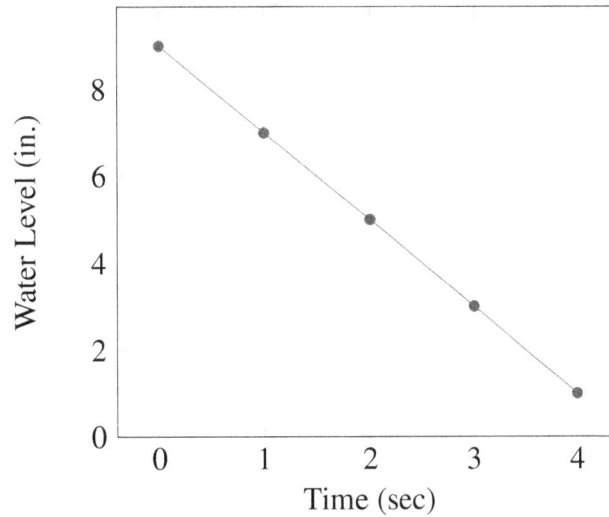

In 1.5 seconds, how many inches is the water level at?

Problem 9.20 A plant that Kelly bought at a gardening store initially stands at 10 cm. tall and on average, it grows 1.5 cm. every year. Approximately, how tall will the plant be in 6 years from the day Kelly bought the plant?

9.3 Practice Questions

Problem 9.21 A group of customer service surveys were sent out at random. The scores were 90, 50, 70, 80, 70, 60, 20, 30, 80, 90, and 20. Find the mean score.

Problem 9.22 Jamie started off the semester with the following six grades on her quizzes:

$$70, 75, 75, 85, 85, 90.$$

What is the minimum number of quizzes Jamie must take to increase her average to at least 90? Assume that the maximum score for any quiz is 100.

Problem 9.23 Suppose you have a list of 4 numbers each chosen from 1 to 10 (inclusive). How many different medians are possible?

Problem 9.24 Suppose you have a list of 9 elements where each element is an nonnegative integer less than or equal to 20. What is the largest possible difference between the median and mean of this list?

Problem 9.25 Given below is a bar graph representing the grade distribution of all students in a different class.

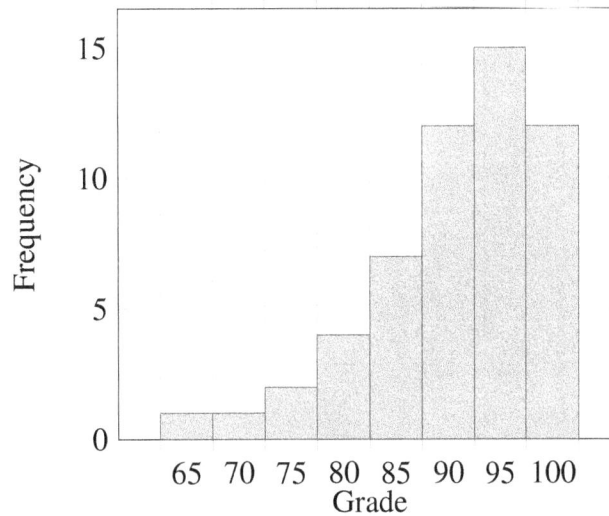

Which of the following is greater: average grade or median grade? Which grade is a better statistic for the overall class performance on the test?

Problem 9.26 Given below is a bar graph representing the ages of all students in a class.

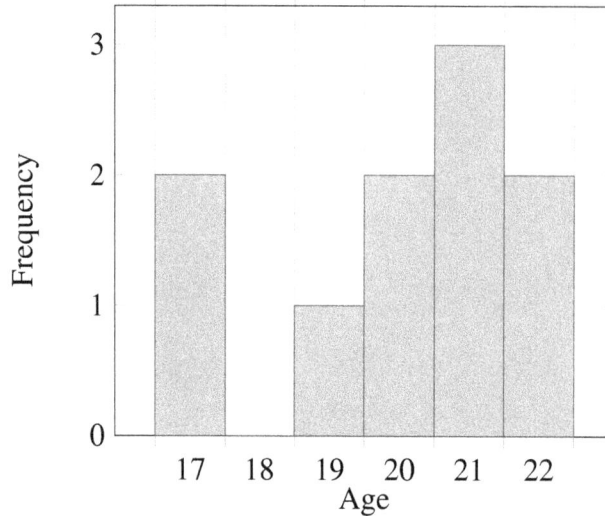

Which age occurs the most in this class?

Problem 9.27 Suppose you have a list. You randomly pick one element from that list. Prove or disprove the following. Hint: Remember to disprove something all you need to do is give an example.

(a) You are most likely to pick the (or one of them if there are multiple) mode of the list.

(b) The probability you pick an element larger than the median is $1/2$.

Problem 9.28 Pretend that everyone likes all the Harry Potter books equally, so every person has an equal chance of picking any of the books as their favorite. Suppose you poll 5 people. What is the probability that exactly 2 people had the 3rd book as their favorite and exactly 2 people had the 4th book as their favorite?

Problem 9.29 The mean width of 12 iPads is 5.2 inches. The mean width of 8 Kindles is 4.7 inches. What is the mean width of the 12 iPads and 8 Kindles?

Problem 9.30 The mean of four numbers is 71.5. If three of the numbers are 58, 76, and 88, what is the value of the fourth number?

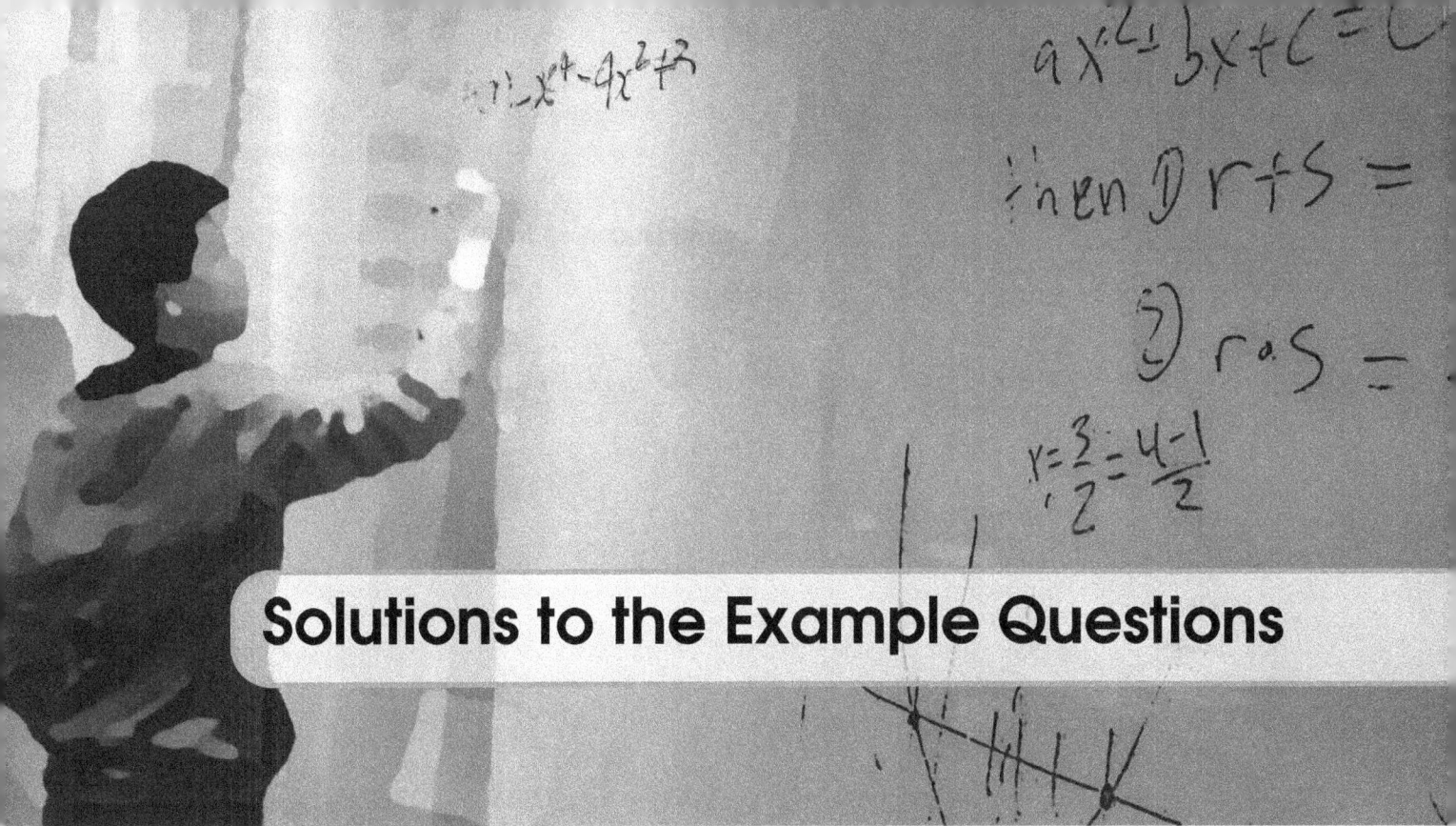

Solutions to the Example Questions

In the sections below you will find solutions to all of the Example Questions contained in this book.

Quick Response and Practice questions are meant to be used for homework, so their answers and solutions are not included. Teachers or math coaches may contact Areteem at info@areteem.org for answer keys and options for purchasing a Teachers' Edition of the course.

1   Solutions to Chapter 1 Examples

Problem 1.1 In the United States, how many state names begin with the letter "M"?

Answer

8

Solution

In alphabetical order, the U.S. states that begin with the letter "M" are Maine, Maryland, Massachusetts, Michigan, Minnesota, Mississippi, Missouri, and Montana. There are 8 U.S. states that begins with the letter "M".

Problem 1.2 Suppose the numbers $0, 1, 2, \ldots, 9$ are represented as on a digital clock as shown below:

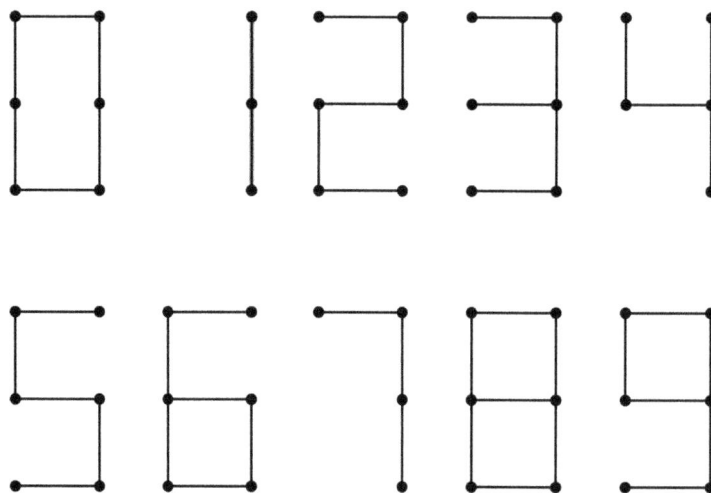

Of these digits, how many numbers have

(a) a horizontal line of symmetry?

Answer

4

Solution

Note that the digits $0, 1, 3$, and 8, have a horizontal line of symmetry as shown below:

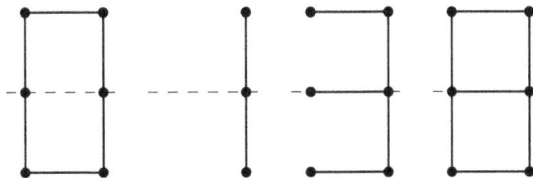

There are 4 digits with a horizontal line of symmetry.

(b) a vertical line of symmetry?

Answer

3

Solution

Note that the digits $0, 1$, and 8, have a vertical line of symmetry as shown below:

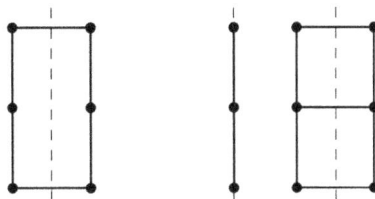

There are 3 digits with a vertical line of symmetry.

(c) both horizontal and vertical lines of symmetry?

Answer

3

Solution

From the previous two questions, we can observe that the only digits satisfying the above conditions are $0, 1$, and 8.

(d) neither horizontal or vertical lines of symmetry?

Answer

6

Solution

From the previous questions, we can deduce that the digits satisfying the neither symmetry condition are $2, 4, 5, 6, 7$, and 9.

Problem 1.3 How many one-digit positive integers are there?

Answer

9

Solution

The numbers are $1, 2, 3, 4, 5, 6, 7, 8, 9$. There are 9 one-digit positive integers.

Problem 1.4 How many 2-digit integers are there?

Answer

90

Solution

The two-digit numbers are $10, 11, 12, \ldots, 99$. It is difficult to count these numbers using your hands so we will adopt a more convenient strategy.

Let's start by counting the number of positive integers in a smaller set of numbers. How many positive integers lie between 10 to 15?

These positive integers are listed below:

$$10, 11, 12, 13, 14, 15$$

There are 6 integers that are listed above. If we consider subtracting the largest number in the list with the smallest number, we observe that the difference is:

$$15 - 10 = 5$$

If we add 1, we obtain the number of positive integers that lie between 10 and 15. This can be extended to positive integers between 10 and 99.

There are a total of

$$99 - 10 + 1 = 90$$

2-digit integers that lie between 10 and 99.

Problem 1.5 Suppose you flip a fair coin three times. How many possible outcomes are there?

Answer

8

Solution

Note that for each toss, you can either obtain heads or tails. Given this, the possible outcomes are:

$$HHH, HHT, HTH, THH, HTT, THT, TTH, TTT$$

There are a total of 8 outcomes.

Problem 1.6 Suppose you flip a coin a total of 3 times.

(a) Make a list of all outcomes with exactly 2 heads.

Solution

The outcomes with exactly 2 heads are:

$$HHT, HTH, THH.$$

(b) Make a list of all outcomes with exactly 2 tails.

Solution

The outcomes with exactly 2 tails are:

$$TTH, THT, HTT.$$

These outcomes can be determined by replacing each H with a T and vice-versa from Part (a).

Problem 1.7 Adam and Bob played ping-pong. They agreed that whoever won 3 games first was the final winner. Assume that Bob was the final winner. In how many ways could the games have been played out?

Answer

10

Solution

We will use brute force to determine the various outcomes of the games. Given that Bob is the final winner, the score of the ping-pong match could either be 3-0, 3-1, or 3-2.

If the final score is 3-0, the possibility is

$$BBB.$$

There is only 1 possibility.

If the final score is 3-1, the possibilities are

$$ABBB, BABB, BBAB.$$

There are 3 possibilities.

If the final score is 3-2, the possibilities are

$$AABBB, ABABB, ABBAB, BAABB, BABAB, BBAAB.$$

There are 6 possibilities.

Therefore, there are

$$6+3+1=10$$

total ways that the game could have been played out.

Problem 1.8 Suppose Adam and Bob have a ping-pong rematch. They agreed to play until one person wins three games *in a row*. Suppose this time it takes 6 games to determine a winner, and the winner is Adam. In how many ways could the games have been played out?

Answer

3

Solution

Since Adam is the winner of the match, the last three games have to be *AAA*. Therefore, the desired outcomes must be of the form

$$- - -AAA$$

The third game must be won by Bob. This is to prevent Adam from winning three games in a row before all six games have been played out. Therefore, the desired outcomes must be of the form

$$- -BAAA$$

Since Bob didn't win the match, Adam must win at least one of the first two games because if Bob wins the first two games, Bob would win the match.

Therefore, the possible outcomes of the games are given as follows:

$$AABAAA, ABBAAA, BABAAA.$$

This yields a total of 3 possibilities.

Problem 1.9 Suppose John has 3 colored shirts (red, white, and blue) and 2 pairs of jeans (light-washed, dark-washed). Each outfit consists of one shirt and one pair of jeans. How many possible outfits can John wear?

Answer

6

Solution

Let R, W, B be the shirts and L, D be the type of jeans that John owns.

Since we need a choice of one shirt and one pair of jeans to create an outfit, the possible outfits that he could wear are

$$RL, WL, BL, RD, WD, BD.$$

There are 6 total outfits that John could wear.

Problem 1.10 Later that day, John decides to get some lunch. He went to his favorite Asian restaurant where customers are allowed to customize their rice plate with a choice between 2 types of rice (white, brown), 3 types of meat (chicken, beef, shrimp), and 2 choices of vegetables (broccoli, string beans). Assuming a rice plate consists of one choice of rice, meat, and vegetables, how many different rice plates can John choose from?

Answer

12

Solution

John is able to choose between white rice W and brown rice B for rice selections, chicken C, beef B and shrimp S for meat selections, and broccoli B and string beans S for vegetable selections.

Therefore, the possible rice plates that John can choose are

$$WCB, WCS, WBB, WBS, WSB, WSS,$$

$$BCB, BCS, BBB, BBS, BSB, \text{and } BSS.$$

There are 12 possible rice plates John can choose from.

2   Solutions to Chapter 2 Examples

Problem 2.1 A restaurant has 5 choices of appetizers, 3 choices for desserts, and for entree they offer 7 types of burgers and 4 types of salads. How many different meals (appetizer, entree, dessert) are available?

Answer

165

Solution

Note that there is a total of

$$4 + 7 = 11$$

entrees to choose from. Therefore, since a person can order one of 5 appetizers, one of 11 entrees, and one of 3 desserts, there are

$$5 \times 11 \times 3 = 165$$

different meals available for the person.

Problem 2.2 Suppose the Martian alphabet has 10 letters, and a Martian word is any sequence of 5 letters, allowing repeats.

(a) How many Martian words are there in total?

Answer

$10^5 = 100000$.

Solution

Since the Martian alphabet has 10 letters, and there are 5 slots to create a Martian word, we observe that there are

$$10 \times 10 \times 10 \times 10 \times 10 = 10^5 = 100000$$

Martian words that can be formed.

(b) How many Martian words are there with no repeating letters?

Answer

30240

Solution

Given the additional condition of having no repeated letters, for the first letter, there are 10 choices. For the second letter, there are 9 choices since the first letter has been already chosen. For the third letter, there are 8 choices since the first and second letters were chosen. This process repeats until all of the letters are filled in the 5−letter word.

Therefore, there are

$$10 \times 9 \times 8 \times 7 \times 6 = 30240$$

Martian words without repeated letters.

(c) How many Martian words are there with at least one pair of repeating letters?

Answer

$100000 - 30240 = 69760$.

Solution

From the previous questions, there are a total of 100000 Martian words, in which 30240 of the Martian words contains no repeated letters. Therefore, the remaining

$$100000 - 30240 = 69760$$

Martian words must contain at least one pair of repeating letters.

Problem 2.3 Suppose that a restaurant sells 7 different burgers and 4 different salads. Two people decide to order different things, but both order a burger or both order a salad. How many different ways can this happen?

Answer

54

Solution

Suppose that the two people order burgers. Then the first person has a choice between 7 burgers. Because the two people are required to order different things, the second person has a choice between 6 burgers. Therefore, there are

$$7 \times 6 = 42$$

different ways for two people to order burgers. Similarly, for salads, the first person has a choice between 4 salads and the second person has a choice between 3 salads. Therefore, there are

$$4 \times 3 = 12$$

different ways for two people to order salads. Therefore, there is a total of

$$42 + 12 = 54$$

different ways for two people to order burgers or salads without ordering the same kind of food.

Problem 2.4 There are 40 lottery balls labeled from 1 to 40. How many ways are there to draw 5 lottery balls, in order one after another, if we replace the ball after each pick? (That is, it is possible to pick the same ball more than once.)

Answer

102400000

Solution

Per lottery ball draw, there are 40 lottery balls to choose from. Since we are drawing the lottery balls 5 times, there are

$$40 \times 40 \times 40 \times 40 \times 40 = 40^5 = 102400000$$

ways to draw 40 lottery balls with replacement.

Problem 2.5 Consider three cities: City A, City B, and City C. There are 5 one-way roads from $A \to B$, 4 one-way roads from $B \to A$, 3 one-way roads from $B \to C$, and 2 one-way roads from $C \to A$ (and no other roads connect the cities). How many loops are there from City A back to City A (without visiting any other city twice)? Note a loop must leave City A, but does not need to visit all the other cities.

Answer

$5 \times 4 + 5 \times 3 \times 2$.

Solution

Note that there are two ways of going from A to A. Namely, one can take the $A \to B \to A$ path or the $A \to B \to C \to A$ path.

For the first case, there are 5 choices of paths to take from A to B. Per choice, there are 4 choices of paths to take from B to A. Therefore, there are

$$5 \times 4 = 20$$

different paths to take without going to C. The second case allows one to go to C which can be achieved in

$$5 \times 3 = 15$$

ways. Since there are 2 paths from C to A, there are

$$15 \times 2 = 30$$

total ways to get from A to A going through C. Therefore, there are

$$20 + 30 = 50$$

possible paths from A to A without visiting any place twice and all roads are one-way.

Problem 2.6 Suppose Tom from Texas is decided where to go on vacation. He can choose to go to Los Angeles, New York, or Chicago. There are 4 airlines from his town in Texas that fly to Los Angeles, 6 that fly to New York, and 8 that fly to Chicago (and the same number of airlines fly back to Texas from each city). How many total round trips are possible?

Answer

116

Solution

Tom from Texas has the option to go to one of three cities, namely, Los Angeles, New York, or Chicago.

If Tom decides to go to Los Angeles for vacation, then he has 4 airline options to transport himself from Texas to Los Angeles. Per airline, he has 4 airline options to choose on his trip back from Los Angeles to Texas. Therefore, there are

$$4 \times 4 = 16$$

possibilities to book a round trip flight from Texas to Los Angeles. Similarly, there are

$$6 \times 6 = 36$$

possibilities to book a round trip flight from Texas to New York and,

$$8 \times 8 = 64$$

possibilities to book a round trip flight from Texas to Chicago. Therefore, there is a total number of

$$16 + 36 + 64 = 116$$

possible round trip flights to choose for his vacation.

Problem 2.7 Suppose a football team has 15 members. The team plays two games, and must choose a captain and co-captain for both games. Suppose that the captain for the first game cannot be a captain or co-captain for the second game (but the first co-captain can be either captain again). How many ways can the team choose a captain and co-captain for the two games?

Answer

38220

Solution

For the first game, we will select a captain and then select a co-captain. Since there are 15 members on the team, there are 15 options for the captain. After the captain is selected, there are 14 options for the co-captain. This implies that for the first game, there are

$$15 \times 14 = 210$$

possibilities to select a captain and co-captain for the first game.

For the second game, we will again select a captain and then select a co-captain. The captain from the first game is not allowed to be the captain or the co-captain, so this implies that there are

$$15 - 1 = 14$$

members to choose for the second game. Using the similar method discussed above, there are

$$14 \times 13 = 182$$

possibilities to select a captain and co-captain for the second game.

Therefore, there are

$$210 \times 182 = 38220$$

possible ways of choosing a captain and a co-captain for the two games.

Problem 2.8 How many ways are there to put a white and black rook on a 8×8 chessboard so that neither can attack the other? (Rooks can only attack along rows and columns but not along the diagonals.)

Answer

$64 \cdot 49 = 3136$.

Solution

Suppose we have an 8×8 board. Allow us to place the white rook on the board first. Since there is nothing currently on the board, we are free to choose any location on the 8×8 board for the white rook. There are 64 spots for the white rook.

Note that for any chosen position, the white rook is threatening $7 + 7 = 14$ squares and occupies 1 square. Therefore, this leaves us with

$$64 - 14 - 1 = 49$$

possible squares for the black rook.

There are

$$64 \times 49 = 3136$$

possible arrangements of two nonthreatening rooks on an 8×8 chessboard.

Problem 2.9 Suppose you put 3 rooks (white, black and gray) on a chessboard (so none of the rooks attack each other). The white rook is in the top row of the board, the black rook is in the bottom row of the board, and the gray rook is in one of the remaining rows. How many ways can this be done?

Answer

2016

Solution

Given that the white rook must be in the top row, there are 8 possible arrangements. The black rook must be on the bottom row of the chessboard. There are 7 possible arrangements for the black rook to be located in a non-threatening position.

For the gray rook, since there are 6 spaces for the 6 inner rows of the chessboard, there are

$$6 \times 6 = 36$$

options for the gray rook. Thus, the number of arrangements that can be made with the above conditions is

$$8 \times 7 \times 36 = 2016.$$

Problem 2.10 In chess, the king can only move one space in any chosen direction. How many ways are there to put a white king and a black king on the chessboard so that they do not attack each other?

Answer

3612.

Solution

Depending on the location of the white king, there are 3 cases to consider: (i) in one of the 4 corners, (ii) on one of the edges (but not in a corner), or (iii) not on any of the edges.

In case (i), there are 4 spaces for the white king. This leaves 3 threatening spaces and 1 occupied space for the black king. There are

$$64 - 4 = 60$$

spaces for the black king to meet the conditions above. This implies that there are

$$4 \times 60 = 240$$

arrangements of the kings for the first case.

In case (ii), since there are 6 spaces per side of the 8×8 chessboard, there are

$$6 \times 4 = 24$$

spaces for the white king. This leaves 5 threatening spaces and 1 occupied space for the black king. There are

$$64 - 6 = 58$$

spaces for the black king to meet the conditions above. This implies that there are

$$24 \times 58 = 1392$$

arrangements of the kings for the second case.

In case (iii), there are 36 remaining spaces for the white king. This leaves 8 threatening spaces and 1 occupied space for the black king. There are

$$64 - 9 = 55$$

spaces for the black king to meet the conditions above. This implies that there are

$$36 \times 55 = 1980$$

arrangements of the kings for the last case.

Therefore, this gives us a total of:

$$240 + 1392 + 1980 = 3612$$

arrangements of two non-attacking kings.

3 Solutions to Chapter 3 Examples

Problem 3.1 There are 40 lottery balls labeled from 1 to 40. How many ways are there to draw 5 lottery balls, in order one after another, if we do not replace the ball after each pick? (That is, it is not possible to pick the same ball more than once.)

Answer

78960960

Solution

There are 40 possible lottery balls to choose for the first drawing. After the first lottery ball is chosen, there are 39 remaining lottery balls to choose for in the second drawing. Continuing the procedure, there is a total of

$$40 \times 39 \times 38 \times 37 \times 36 = 78960960$$

ways to choose 5 lottery balls without replacement.

Problem 3.2 There are 40 lottery balls labeled from 1 to 40. How many ways are there to draw 5 lottery balls all at once? (That is, it is not possible to draw the same ball twice, and the 5 balls are in no particular order.)

Answer

$$\binom{40}{5} = \frac{40!}{5!35!} = 658008.$$

Solution

This is equivalent to drawing 5 indistinguishable lottery balls from a pool of 40 balls. From the previous problem, there is a total of

$$40 \times 39 \times 38 \times 37 \times 36 = 78960960$$

ways to choose 5 lottery balls without replacement.

Since the lottery balls are indistinguishable, we divide the number of arrangements of the 5 balls to remove the duplicates from the above count. Therefore, there are

$$78960960 \div 5! = 658008$$

ways to draw 5 lottery balls all at once.

Problem 3.3 Suppose five people are to be seated in a row of 9 chairs. How many possible seating arrangements can be made if there must be a seat in between each person?

Answer

120

Solution

Given the conditions of the problem, the 5 people must sit in seats $1, 3, 5, 7, 9$.

Since these seats are required to be filled, we determine the number of ways to permute 5 people. This is done in

$$5 \times 4 \times 3 \times 2 \times 1 = 120$$

ways.

Problem 3.4 Suppose you have a club of 20 members. The club chooses four officers: President, Vice-President, Treasurer, and Secretary. They also choose someone to be in charge of fundraising. The four officers must all be different, but the member in charge of fundraising can be one of the officers. How many ways can we choose the four officers in the club with no restrictions?

Answer

2325600

Solution

We can first choose the President in the club. There are 20 options.

After the President is chosen, when choosing the Vice-President, there are 19 options.

We repeat this process until all of the positions are filled.

Therefore, there are

$$20 \times 19 \times 18 \times 17 = 116280$$

ways to fill the four positions.

Additionally, we wish to appoint someone in the club to take charge of fundraising. This can be any one of the 20 members in the club. Therefore, we have:

$$116280 \times 20 = 2325600$$

ways to appoint four officers and a person in charge of fundraising.

Problem 3.5 Suppose a football team has 10 members. How many ways are there to choose 2 co-captains?

Answer

$$\frac{10 \cdot 9}{2} = 45.$$

Solution

Firstly, suppose that there is an order imposed on the selection of the co-captains. There are 10 options for the first co-captain and 9 options for the second co-captain (the first member excluded from being chosen as the first co-captain). This yields a total of

$$10 \times 9 = 90$$

possible ways to achieve this. However, since we do not care about the order of the choices of the co-captains, we divide by the number of possible arrangements of 2 co-captains.

Therefore, the answer is:

$$\frac{10 \times 9}{2} = 45$$

ways to select two co-captains from a team of 10 members.

Problem 3.6 Eight friends all live in the same dorm. They receive 3 movie tickets for the weekend. They decided to draw the tickets randomly. If all the tickets are identical and a friend gets at most one ticket, how many ways can the tickets be distributed?

Answer

56

Solution

Let A, B, C, D, E, F, G, H indicate the 8 friends. If we allow the ordering of the distribution of the tickets to matter, then the tickets can be distributed in

$$8 \times 7 \times 6 = 336$$

ways. If the tickets are identical, order does not matter and the pre-determined 336 ways contain duplicate arrangements. To rid these duplicates, we divide by the number of ways to arrange 3 tickets. This is achieved in

$$3 \times 2 \times 1 = 3! = 6$$

ways. Therefore, the number of ways that the tickets can be distributed is

$$\binom{8}{3} = \frac{8!}{3! \times 5!} = \frac{8 \times 7 \times 6}{3!} = 56.$$

Problem 3.7 Suppose you flip a coin a total of 9 times.

(a) How many total outcomes are there?

Answer

$2^9 = 512$

Solution

Since each coin toss yields a possibility of either heads or tails and that the coin is tossed 9 times, there are

$$2^9 = 512$$

total outcomes.

(b) How many outcomes have exactly 3 heads?

Answer

84

Solution

In 9 tosses, we require 3 occurrences of getting heads and 6 occurrences of getting tails. This can be denoted with 3 H's and 6 T's.

Every possible arrangement of 3 H's and 6 T's corresponds to the number of outcomes with exactly 3 heads. The number of ways to do this is:

$$\binom{9}{3} = \frac{9!}{3! \times 6!} = 84.$$

(c) How many outcomes have more heads than tails? Hint: Compare this to the number of outcomes that have more tails than heads?

Answer

$2^8 = 256$

Solution

Note that there is no outcome that contains the same amount of heads and tails.

Furthermore, note the number of outcomes that have more heads is equal to the number of outcomes with more tails. This can be observed if we consider one outcome and replace every H with T and vice versa.

Therefore, there are

$$\frac{1}{2} \times 2^9 = 2^8 = 256$$

outcomes with more heads than tails.

Problem 3.8 Suppose below is a map of a city you want to travel from A to B.

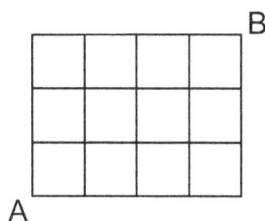

(a) If each square in the diagram is a square "block" what is the minimum number of blocks it takes to get from A to B?

Answer

7

Solution

In the diagram above, we are required to travel 3 blocks up and 4 blocks right to get from A to B. Therefore, the minimum number of blocks it takes to get from A to B is

$$3 + 4 = 7$$

.

(b) How many paths shortest length are there from A to B?

Answer

35

Solution

Let R stand for "right" and U stand for "up". Note that any one of the shortest paths consists of 3 blocks up (3 U's) and 4 blocks right (4 R's).

A unique path is determined by the arrangement of 3 U's and 4 R's. The number of possible ways to arrange 3 U's and 4 R's is

$$\binom{7}{3} = \frac{7!}{3! \times 4!} = 35.$$

Problem 3.9 How many ways can you put 8 identical mutually non-attacking rooks on a 8×8 chessboard?

Answer

$8! = 40320.$

Solution

This problem can be interpreted in multiple ways. In our solution, we will discuss two methods.

Firstly, we select one row out of 8 to place a rook in. Since each row has 8 available spaces, this can be done in 8 ways.

After that rook is placed, we attempt to place a second rook in another selected row. There are now 7 available spaces since placing the second rook in the corresponding space to the first rook will make the rooks attack each other.

Continuing this process, we observe that this can be done in

$$8! = 8 \times 7 \times 6 \times 5 \times 4 \times 3 \times 2 \times 1 = 40320$$

ways.

Alternatively, we can approach this solution from another perspective. We can start off by placing the first rook in one of 8^2 spaces. In doing so, this removes a row and column from the spaces available for the next rook.

The next rook can be placed in 7^2 spaces. Continuing this process, we observe that the rooks can be placed in

$$8^2 \times 7^2 \times 6^2 \times 5^2 \times 4^2 \times 3^2 \times 2^2 \times 1^2 = 1625702400$$

ways.

However, this accounts for duplicate arrangements since the rooks are indistinguishable. We have to divide the above result by 8! to rid the duplicates. Therefore, there are

$$1625702400 \div 8! = 40320$$

ways to arrange 8 identical mutually non-attacking rooks on a 8×8 chessboard.

Problem 3.10 10 points are marked on the plane. How many different triangles can be formed using these points as vertices if

(a) no three of the points are in a straight line?

Answer

120

Solution

Note that if no three points are on a straight line we can choose any group of 3 points to form a triangle. This can be done in

$$\binom{10}{3} = 120$$

ways.

(b) 5 points are on one line, and the other 5 points are on another line parallel to the first?

Answer

100

Solution

To form a triangle in the setting above, we are required to choose 2 points on one line and 1 point on the other line. Respectively, this can be done in

$$\binom{5}{2} = 10$$

and

$$\binom{5}{1} = 5$$

ways. Since there are two lines, we have to choose one of the lines to contain two points, which is done in

$$\binom{2}{1} = 2$$

ways. Therefore, the total number of ways to achieve this is

$$2 \times 5 \times 10 = 100.$$

4   Solutions to Chapter 4 Examples

Problem 4.1 How many rearrangements can be made of the letters in the following words?

(a) STOP

Answer

24

Solution

There are 4 letters to choose in the first letter slot. After the first letter is chosen, there are 3 remaining letters to choose in the second letter slot.

Repeating this argument for all 4 letters yields that the number of rearrangements that can be made using the letters of $STOP$ is

$$4! = 4 \times 3 \times 2 \times 1 = 24.$$

(b) RHYTHM

Answer

360

Solution

Pretend that all of the letters are distinct. There are

$$6! = 6 \times 5 \times 4 \times 3 \times 2 \times 1 = 120$$

ways to rearrange 6 letters to form distinct words.

Since there are 2 H's in $RHYTHM$, we need to rid the duplicates by dividing the total number of ways to rearrange 6 different letters by $2! = 2$.

Therefore, the answer is

$$\frac{6!}{2!} = \frac{6 \times 5 \times 4 \times 3 \times 2 \times 1}{2 \times 1} = 360.$$

(c) BANANAS

Answer

420

Solution

Pretend that all of the letters are distinct. There are

$$7! = 7 \times 6 \times 5 \times 4 \times 3 \times 2 \times 1 = 840$$

ways to rearrange 7 letters to form distinct words.

Since there are 3 A's and 2 N's in *BANANAS*, we need to rid the duplicates by dividing the total number of ways to rearrange 7 different letters by $3! \times 2! = 6 \times 2 = 12$.

Therefore, the answer is

$$\frac{7!}{3! \cdot 2!} = \frac{7 \times 6 \times 5 \times 4 \times 3 \times 2 \times 1}{3 \times 2 \times 1 \times 2 \times 1} = 420.$$

Problem 4.2 There are 4 terminals in an airport with each consisting of 5 flight attendents. An airline is asked to select a group of 4 flight attendents chosen in the following way: randomly choose 2 terminals and per terminal, randomly choose 2 flight attendents. How many ways can this be done?

Answer

600

Solution

Start off by choosing the 2 terminals from the 4. This can be achieved in

$$\binom{4}{2} = 6$$

ways.

For every choice of 2 terminals, there are

$$\binom{5}{2}\binom{5}{2} = 100$$

ways of choosing the flight attendents.

Therefore, the total number of ways to choose a group of 4 flight attendents is equal to

$$\binom{4}{2}\binom{5}{2}\binom{5}{2} = 600$$

Problem 4.3 Suppose you have 8 people and you want to take a photo of everyone lined up.

(a) How many different photos are there in total?

Answer

40320

Solution

In the line, there are 8 choices for the first spot. After the first choice was made, there are 7 choices for the second spot.

Repeating this yields that the total number of different photos that can be taken is

$$8! = 8 \times 7 \times 6 \times 5 \times 4 \times 3 \times 2 \times 1 = 40320.$$

(b) How many photos are there if two of the people are identical twins (who are also dressed the same)?

Answer

20160

Solution

Using the previous solution, we know that there are 40320 ways to arrange 8 distinguishable people from each other.

However, we do not care about the order of the twins because we cannot tell them apart in the photo.

Therefore, we need to remove the duplicate photos by dividing it by the number of ways to arrange 2 distinguishable people. This can be done in $2! = 2$ ways.

The answer is
$$\frac{8!}{2!} = 20160.$$

(c) How many photos are there if there is a couple who wants to stand together?

Answer

10080

Solution

Note that there are $2 = 2!$ ways to decide the arrangement of the couple.

If we then treat them as a single 'object' we must arrange 7 'objects'. This can be done in

$$7! = 7 \times 6 \times 5 \times 4 \times 3 \times 2 \times 1 = 5040$$

ways. Therefore, the answer is

$$2! \times 7! = 2 \times 5040 = 10080.$$

Problem 4.4 One student has 6 books and another student has 8.

(a) How many ways can they trade 3 books of the first student for 3 books of the second?

Answer

1120

Solution

In order to trade 3 books, the first student must select 3 books from the 6 books he has. The second student must select 3 books from the 8 books. Note that in the selection process, the order of how the books are chosen does not matter.

The first student can do this in
$$\binom{6}{3} = 20$$
ways and the second student can do this in
$$\binom{8}{3} = 56$$
ways. Therefore, the total number of ways that this can be done is:
$$\binom{6}{3} \times \binom{8}{3} = 20 \times 56 = 1120.$$

(b) Suppose now the first student gives 3 of their books to the second, and then the second gives 3 books to the first. Unfortunately, the second student has a very bad memory, so they maybe give the first student some of their own books back. How many ways can the "trade" now take place?

Answer

3300

Solution

In order to trade 3 books, the first student must select 3 books from the 6 books he has. Since the second student has poor memory, he must select 3 books from the $8 + 3 = 11$

books he had acquired. Note that in the selection process, the order of how the books are chosen does not matter.

The first student can do this in

$$\binom{6}{3} = 20$$

ways and the second student can do this in

$$\binom{11}{3} = 165$$

ways. Therefore, the total number of ways that this can be done is:

$$\binom{6}{3} \times \binom{11}{3} = 20 \times 165 = 3300.$$

Problem 4.5 Suppose you have 5 men and 5 women at a dance class. How many ways are there to divide the 10 into 5 pairs if

(a) each pair is a male and a female?

Answer

$5! = 120.$

Solution

If we line the men up in a row and let the girls choose their partners, the first girl would have to select one of 5 men to choose from. After a man was chosen, the second girl would have to select one of 4 men to choose from.

Repeating this process, we have

$$5! = 5 \times 4 \times 3 \times 2 \times 1 = 120$$

ways that this can be done.

(b) there are no restrictions in how the pairs are chosen?

Answer

$$\frac{10!}{(2!)^5 \times 5!} = 945$$

Solution

Arrange everyone into pairs by lining them up. We then do not care about the arrangement of the pairs in a line (which can be done in $(2!)^5$ ways). Finally, we do not care about the order of the pairs, so we divide by 5!.

Problem 4.6 James, Jim, and 6 other friends line up for a photograph. James and Jim do not stand next to each other. How many different photographs are possible?

Answer

30240

Solution 1

Use Complementary Counting: There are 8! total photographs of all 8 friends in a line. If we group James and Jim together, there are $2! \times (6+1)!$ photographs with James and Jim next to each other. Hence there are

$$8! - 2! \times 7! = 30240$$

photographs with James and Jim separated.

Solution 2

If James is in the first spot, there are 6 places for Jim. If James is in the second spot there are 5 positions for the Jim. Continuing with the cases, we see that there are

$$2 \times 6 + 6 \times 5 = 42$$

ways to arrange James and Jim. Since there are 6! ways to arrange the other friends, in total there are

$$42 \times 6! = 30240$$

different photographs.

Problem 4.7 Suppose you have 10 numbered balls and 5 numbered boxes. How many ways are there to put the balls into the boxes if:

(a) there are no restrictions?

Answer

9765625

Solution

Consider the first ball. There are 5 boxes to place the first ball in.

Similarly, there are also 5 boxes to place the second ball in. Repeating this process yields that the number of possible ways to do this is:

$$5^{10} = 9765625.$$

(b) no box has more than 2 balls?

Answer

113400

Solution

By the construct of the problem, each box must contain exactly 2 balls. The only way to achieve this is by dividing the balls into 5 pairs. There are 10 options for the first ball in the first pair and 9 options for the second ball in the first pair.

We repeat this method of counting to determine that the number of ways that a box has no more than 2 balls is

$$\frac{10!}{(2!)^5} = \binom{10}{2} \times \binom{8}{2} \times \binom{6}{2} \times \binom{4}{2} \times \binom{2}{2} = 113400.$$

(c) no box has more than 9 balls?

Answer

9765620

Solution

Note that there are only 5 outcomes with 10 balls in a box. Using complementary counting, we subtract this from the total in part (a). to determine that the solution is

$$5^{10} - 5 = 9765625 - 5 = 9765620.$$

Problem 4.8 How many rearrangements of the word *MACHINES* have no vowels next to each other?

Answer

14400.

Solution

The 5 consonants create $5 + 1 = 6$ 'spaces' that can be used for placing the vowels. Hence there are

$$5! \times 6 \times 5 \times 4 = 14400$$

ways to arrange the letters.

Problem 4.9 Suppose a pizza place has 5 toppings available. You want to order 2 different 3-topping pizzas. Suppose repeated toppings are not allowed on a single pizza, and the order of the toppings on a pizza does not matter. If you only care which two pizzas you get, how many ways are there to make the order?

Answer

45

Solution

Since the pizza has 5 toppings made available and each pizza requires 3 nonrepeated toppings, there are

$$\binom{5}{3} = 10$$

combinations of toppings that are made available for pizzas. Of the 10 possible pizzas, we wish to choose 2 pizzas for our order (without order), which can be done in

$$\binom{10}{2} = 45$$

ways.

Problem 4.10 Suppose you have a group of 8 people. How many different photographs are there of everyone lined up if 2 of the people are identical twins and 3 of the people are identical triplets (the twins and triplets dress identically)?

Solution

$$\frac{8!}{3! \cdot 2!} = \binom{8}{3} \cdot \binom{5}{2} \cdot 3! = 3360.$$

5 Solutions to Chapter 5 Examples

Problem 5.1 Suppose that every time you flip a coin, you take one step forward (F) if you get heads and one step backward (B) if you get tails. You can represent the result by a string of letters like "FFBFBBF...." How many ways can you:

(a) take 10 steps?

Answer

$2^{10} = 1024$.

Solution

You must count all possible strings of length 10.

To take 10 steps, we require 10 coin flips. Each coin flip has 2 outcomes so there is a total of

$$2^{10} = 1024$$

outcomes. Each outcome represents a different sequence of steps. Therefore, there are 1024 different ways to take 10 steps.

(b) take 10 steps and wind up where you started?

Answer

$\binom{10}{5} = 252$.

Solution

In order to generate a path where you end up at your initial position after 10 steps, we require the same number of forward and backward steps.

Since we have 10 steps, we need 5 F's and 5 B's. Arranging 5 F's and 5 B's can be done in

$$\binom{10}{5} = 252$$

ways.

Problem 5.2 Suppose you have 3 identical balls and 3 (different) boxes.

(a) How many different ways can you put the balls into the boxes? List all the outcomes using "stars and bars".

Answer

$$\binom{3+3-1}{3} = \binom{5}{3} = 10.$$

Solution

We have 3 $*$'s (stars) and 2 $|$'s (bars) $***||, |***|, ||***, **|*|, *|**|, **||*, *|**, |*$ $*|*, |*|**, *|*|*$.

(b) Explain how to approach part (a) using cases (assuming you didn't know "stars and bars").

Answer

$3 + 3 \cdot 2 + 1 = 10.$

Solution

Break into cases based on whether we have (i) 3 balls in a single box, (ii) 2 balls in one box and one in another, (iii) 1 ball in each box.

(c) If the first box must have at least one ball how many outcomes are there?

Answer

$$\binom{2+3-1}{2} = \binom{4}{2} = 6.$$

Solution

The first box must contain at least one ball. The remaining 2 balls can be distributed freely among any of the 3 boxes.

Alternatively, we can subtract the $\binom{3+2-1}{3} = \binom{4}{3} = 4$ ways to distribute the 3 balls among the second and third boxes from the total: $10 - 4 = 6$.

Problem 5.3 Suppose a cookie store sells 4 types of cookies: chocolate, peanut butter, ginger, and sugar. You want to buy a total of 8 cookies. Suppose the order you buy the cookies does not matter.

(a) How many ways are there to choose the cookies in total?

Answer

165

Solution

Let the 8 cookies represent the "stars". Because we wish to divide the "stars" into 4 types of cookies, we need $4 - 1 = 3$ "bars" to do this.

Therefore, there are

$$\binom{8+4-1}{8} = \binom{11}{8} = 165$$

ways to choose the cookies in total.

(b) How many ways are there to choose the cookies if you must buy at least one of each type?

Answer

$$\binom{4+4-1}{4} = \binom{7}{4} = 35.$$

Solution

Since we are required to have at least one cookie of each type, we reserve 4 of the 8 bought cookies to be one of each type. This leaves us with 4 remaining cookies free of choice.

Let the 4 remaining cookies represent the "stars". Because we wish to divide the "stars" into 4 types of cookies, we need $4 - 1 = 3$ "bars" to do this.

Therefore, there are

$$\binom{4+4-1}{4} = \binom{7}{4} = 35$$

ways to choose the cookies in total.

Problem 5.4 Suppose you have 10 identical balls and 5 numbered boxes. How many ways are there to put the balls into the boxes if:

(a) there are no restrictions?

Answer

1001

Solution

Let the 10 balls represent the "stars" and let the $5 - 1$ "bars" divide the stars into 5 groups, with each group indicating the number of balls in each box.

Therefore, the number of ways to arrange the stars and bars is equal to

$$\binom{10+5-1}{10} = 1001.$$

(b) each box has at least one ball?

Answer

126

Solution

Since we are required to have at least one ball in each box, we reserve 5 of the 10 balls to be placed in each box. From there, it becomes the standard "stars and bars" problem.

Let the 5 balls represent the "stars" and let the $5 - 1$ "bars" divide the stars into 5 groups, with each group indicating the number of balls in each box.

Therefore, the number of ways to arrange the stars and bars is equal to

$$\binom{5+5-1}{5} = 126.$$

(c) no box has more than 2 balls?

Answer

1.

Solution

The only possibility for this to occur is if we equidistribute the 10 balls in 5 boxes. Namely, each box has exactly 2 balls.

Assuming that the balls are indistinguishable, there is only 1 way of doing this.

(d) no box has more than 9 balls?

Answer

996.

Solution

Recall from the previous problem that there are 1001 total ways to arrange the 10 balls in 5 boxes.

Since we require that no box have more than 9 balls, we use complementary counting to remove the cases when a box has more than 9 balls (i.e. when a box has 10 balls). There are 5 ways to put all 10 balls in one of 5 boxes.

Therefore, the number of ways to place 10 balls in 5 boxes such that each box has no more than 9 balls is

$$1001 - 5 = 996.$$

Problem 5.5 Suppose that a person has 10 friends. He owns a movie theater (so has access to as many tickets as he wants) and wants to invite some of his friends to a movie Friday night. How many ways can he give his friends tickets if:

(a) he chooses 6 friends to give one ticket to?

Answer

$$\binom{10}{6} = 210.$$

Solution

This is a combination.

(b) he chooses any number of friends (including none or all) to give one ticket to?

Answer

$2^{10} = 1024.$

Solution

He either gives each friend a ticket or not.

(c) he gives out 10 tickets in total, but it is possible that some friends get more than one ticket (so they can bring their family, or other friends, etc.)?

Answer

$$\binom{10+10-1}{10} = 92378.$$

Solution

We are now putting 10 identical balls (tickets) into 10 boxes (friends) so we use stars and bars.

Problem 5.6 Let $a, b, c, d \geq 0$ be integers. How many solutions to $a+b+c+d = 15$ are there

(a) in total?

Answer

816

Solution

Let the "stars" represent balls and the $4-1$ "bars" represent the division of the balls into 4 groups so that the number of balls in the first group represent the value of a, and so on.

Therefore, the number of ways to arrange the stars and bars is

$$\binom{15+4-1}{15} = 816.$$

(b) with $a \geq 4$?

Answer

364

Solution

To account for the added condition of $a \geq 4$, we reserve 4 of the 15 balls to be placed in the first box. This leaves us with 11 balls free to be arranged in any way.

Applying a very similar method to arrange the remaining 11 balls in 4 boxes, the number of solutions under the given condition is
$$\binom{11+4-1}{11} = 364.$$

(c) with $a = 4$?

Answer

78

Solution

To account for the added condition of $a = 4$, we reserve 4 of the 15 balls to be placed in the first box. This leaves us with 11 balls free to be arranged in any way.

However, since we are required to have 4 balls in the first box, the first bar is fixed. Therefore, we can only arrange the remaining 2 bars.

Applying a very similar method to arrange the remaining 11 balls in 3 boxes, the number of solutions under the given condition is
$$\binom{11+3-1}{11} = 78.$$

Problem 5.7 How many 4 digit numbers that do not contain the digit 0 are there that

(a) are there in total?

Answer

$9^4 = 6561.$

Solution

Every digit can be any besides 0.

(b) have no repeated digits?

Answer

$$9 \cdot 8 \cdot 7 \cdot 6 = \frac{9!}{5!} = 3024.$$

Solution

This is a permutation.

(c) have digits that sum up to 8?

Answer

$$\binom{4+4-1}{4} = 35.$$

Solution

Each digit must be at least 1, use stars and bars.

Problem 5.8 You're playing a game of Farkel using six dice.

(a) How many different possible outcomes are there in rolling six dice?

Answer

462

Solution

Let the "stars" represent the dice and the $6 - 1$ "bars" represent the division of the dice into 6 groups so that the number of balls in the first group represent the dice with value 1, and so on.

Therefore, the number of ways to arrange the stars and bars is

$$\binom{6+6-1}{6} = \binom{11}{6} = 462.$$

(b) How many ways are there to get exactly three 6's?

Answer

35

Solution

Let the "stars" represent the dice and the $6 - 1$ "bars" represent the division of the dice into 6 groups so that the number of balls in the first group represent the dice with value 1, and so on.

Since we are required to put 3 of the balls into the sixth box, the 5th bar and 3 of 6 stars are fixed. Therefore, the number of ways to arrange the stars and bars under this condition is:

$$\binom{3+5-1}{3} = \binom{7}{3} = 35.$$

Problem 5.9 How many integer solutions to the equation $a+b+c = 15$ are there if:

(a) $a,b,c \geq 0$?

Answer

$$\binom{15+3-1}{15} = 136.$$

Solution

Think of any integer as a box made up of 1's. We are then putting 15 1's into 3 boxes.

(b) $a,b,c \geq 1$?

Answer

$$\binom{12+3-1}{12} = \binom{14}{12} = \binom{14}{2} = 91.$$

Solution

Each number must contain at least a 1, so we are left with 12 1's and 3 boxes.

Problem 5.10 How many solutions to $(a+b) \cdot (c+d) = 15$ and $a, b, c, d \geq 0$ are there in total?

Answer

112

Solution

We can use Stars and Bars to solve this problem. Note that there are two ways to factor 15; namely,

$$15 = 1 \times 15 = 3 \times 5.$$

Because we are interested in re-expressing factors of 15 into a sum of two nonnegative integers, we require $2 - 1$ bars to divide stars into 2 groups.

Therefore, the number of solutions to $(a+b) \times (c+d) = 15$ and $a, b, c, d \geq 0$ is:

$$2\binom{15+2-1}{15}\binom{1+2-1}{1} + 2\binom{5+2-1}{5}\binom{3+2-1}{3} = 112.$$

We need to multiply the entire expression by 2 to account for communitivity of multiplication.

6   Solutions to Chapter 6 Examples

Problem 6.1 You have five friends: two are boys and three are girls. You are blind-folded and randomly select two friends (without regard to order).

(a) Write out the list of all possible ways that this can be done. How many ways are there?

Solution

Let the girls be named A, B, C and the boys be named D, F. Then, the possible ways to do this are:
$$AB, AC, AD, AF, BC, BD, BF, CD, CF, DF$$

There are
$$\binom{5}{2} = 10$$
outcomes in the sample space.

(b) What is the list of all possible outcomes that you pick two girls? How many ways are there?

Solution

From the previous solution, we named the girls A, B, C. Therefore, the possible ways to do this are given in the sample space below:
$$AB, AC, BC$$

There are
$$\binom{3}{2} = 3$$
outcomes listed above.

Problem 6.2 Cameron wanted to play a card game with a friend using a standard 52-card deck. He offers his friend \$1 if he drew an ace and \$2 if he drew a spade. How many ways can Cameron's friend earn money?

Answer

16

Solution

For Cameron's friend to earn money, he must draw a spade or an ace from the deck of cards. There are 13 cards with the suit spades and 4 cards with rank ace. However, there is one card that have the rank ace and suit spades (the Ace of Spades). Therefore, the number of cards that either have rank ace or suit spades is

$$13 + 4 - 1 = 16.$$

Therefore there are 16 ways for Cameron's friend to earn money.

Problem 6.3 Students in Areteem Institute were asked which pets (dogs or cats) do they have. In a survey of 100 students, 10 of them answered "No pets", 70 answered "a dog" and 50 answered "a cat". How many students have both a cat and dog?

Answer

30

Solution

Given 100 students, since 10 of the students do not own any pets,

$$100 - 10 = 90$$

students owns at least one pet. Of the 90 remaining students, we wish to distribute the students to one of three regions in the following Venn Diagram:

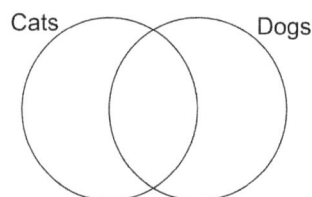

It is given that there are 50 students that answered "a cat". This indicates that the shaded regions shown below must add to 50.

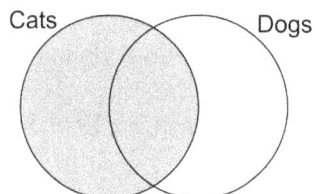

It is also given that there are 70 students that answered "a dog". This indicates that the shaded regions shown below must add to 70.

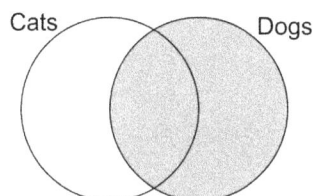

Note that if we add shaded regions above, the overlap of the regions is counted twice. This suggests that the number of students who have at least one cat and one dog is

$$70 + 50 - 90 = 30.$$

This is because the overlap of the number of students represents the number of students that own both cats and dogs. Therefore, there are 30 students that have both a cat and dog.

Problem 6.4 How many numbers less than 100 are divisible by 5 but not 3?

Answer

13

Solution

They are
$$5, 10, 20, 25, 35, 40, 50, 55, 65, 70, 80, 85, 95.$$

Note there are 19 multiples of 5 and 6 multiples of 15, so there are

$$19 - 6 = 13$$

multiples of 5 that are not multiples of 3.

Problem 6.5 Suppose you roll 2 four-sided dice. Let A be the event that the first die is a 4, and B the event that the sum of the two rolls is 6.

(a) Write out the sample space representing a list of all possible outcomes of rolling 2 four-sided dice. How many outcomes are possible?

Solution

The list of all possible outcomes from rolling 2 four-sided dice is given below:

$$(1,1),(1,2),(1,3),(1,4),(2,1),(2,2),\dots(4,4).$$

There are 16 total outcomes.

(b) List the outcomes in event A.

Answer

$(4,1),(4,2),(4,3),(4,4)$

Solution

The possible outcomes are

$$(4,1),(4,2),(4,3),(4,4)$$

(c) List the outcomes in event B.

Answer

$(2,4),(3,3),(4,2)$

Solution

The possible outcomes are

$$(2,4),(3,3),(4,2)$$

(d) List the outcomes in event $A \cap B$.

Answer

$(4,2)$

Solution

The only possible outcome is

$$(4, 2)$$

Problem 6.6 Suppose you roll a six-sided die.

(a) What are the outcomes of the events A, B, and C, if A represents the event when you roll a 3 on a six-sided dice, B represents the event when you roll an odd number on a six-sided dice, and C represents the event when you roll a number greater than 3 on a six-sided dice?

Answer

$A = \{3\}; B = \{1, 3, 5\}; C = \{5, 6\}$.

(b) What are the events: A^c; B^c; $A^c \cap B$; $A \cup B^c$; $(A \cup B^c)^c$?

Answer

$A^c = \{1, 2, 4, 5, 6\}$; $B^c = \{2, 4, 6\}$; $A^c \cap B = \{1, 5\}$; $A \cup B^c = \{2, 3, 4, 6\}$; $(A \cup B^c)^c = \{1, 5\}$.

Problem 6.7 Suppose we are given a bag of 30 balls and 20 cubes. It is known that of the 50 objects in the bag, 40 of them are red and 10 of them are blue. If at least one of the objects in the bag is a blue cube, what is the minimum number of red balls in the bag?

Answer

21

Solution

We note that the bag contains objects definied by its shape (cubes vs. balls) and color (red vs. blue).

Shape and color are independent from each other, so one can interpret the complement of cubed-shaped objects to be ball-shaped objects and the complement of red objects to be blue.

Since we are interested in minimizing the number of red balls in the bag, we want to maximize the number of red cubes in the bag.

Given that there are at least one blue cube in the bag, the maximum number of red cubes is

$$20 - 1 = 19.$$

Since there are 40 total red objects, the minimum number of red balls is

$$40 - 19 = 21.$$

Problem 6.8 Among 50 adults who like either coffee or tea, 30 of them like coffee and 30 of them like tea. A brand new coffee shop managed to increase the number of adults who like coffee by 20%. This decreases the number of adults that strictly love tea by a certain percentage. What is this percentage?

Answer

30%

Solution

Given that of the 50 adults, 30 like coffee and 30 like tea, the overlap represent the number of adults that like both coffee and tea. There are

$$(30 + 30) - 50 = 10$$

adults that like both coffee and tea.

With the introduction of the new coffee shop, this increases the number of adults that like coffee by 20%. Namely, there are

$$30 \times 0.20 = 6$$

additional people that like coffee. This converts 6 people in the tea-only category to people in both coffee and tea category. There are

$$30 - 10 = 20$$

adults in the tea-only category. Therefore, the percent decrease of the number of adults in the tea-only category is

$$\frac{6}{20} = 30\%.$$

Problem 6.9 300 participants of the survey are asked whether they like Coke or Pepsi. According to the survey, $\frac{2}{3}$ participants like Coke and $\frac{1}{2}$ of the participants like Pepsi. Assume no one dislikes both Coke and Pepsi. How many participants like both Coke and Pepsi?

Answer

60

Solution

We know that $\frac{1}{2}$ of the participants like Coke, so a total of

$$\frac{2}{3} \times 300 = 200$$

like Coke. Similarly

$$\frac{1}{2} \times 300 = 150$$

like Pepsi. Hence a total of

$$200 + 150 = 350$$

people like Coke or Pepsi (or both). Since there are 300 people in total,

$$350 - 300 = 50$$

must like both Coke and Pepsi.

Problem 6.10 Sarah is trying to eat healthier. To do so, each day she keeps track of whether she (i) ate a salad, (ii) ate dessert.

(a) For a single day, how many different outcomes are there?

Answer

4

Solution

Note for both eating a salad and eating dessert there are two options: yes or no. Hence there are

$$2 \times 2 = 4$$

outcomes for each day.

(b) For the entire week, how many outcomes are there? What might be a good way for Sarah to keep organize her data?

Answer

$4^7 = 16384$

Solution

There are 4 outcomes each day. This can be visualized using a Venn Diagram with two sets: A for eating a salad, and B for eating dessert. Since there are 7 days in the week (and 4 choices each day), in total there are

$$4^7 = 16384$$

outcomes.

7 Solutions to Chapter 7 Examples

Problem 7.1 For each of the following descriptions of a sequence: (i) write out the first few terms of the sequence, (ii) state whether it is an infinite or finite sequence, (iii) if it is a finite sequence, state its length.

(a) The number of days in the month, starting with January, February, etc.

Answer

$31, 28, 31, 30, \ldots$

Solution

Since January has 31 days, the first term of the sequence is 31. Since February has 28 days, the second term of the sequence is 28. We continue the sequence as follows:

$$31, 28, 31, 30, \ldots.$$

Since there are finitely many months in a year, the sequence is finite with length 12.

(b) The sequence with where the nth term is $4n^2 + 2$.

Answer

$2, 6, 18, 38, \ldots$

Solution

The first few terms of the sequence is

$$2, 6, 18, 38, \ldots$$

Since there is no restriction on the value of n, the sequence is infinite.

(c) The number of ways to invite a group of 0, 1, 2, etc. friends to a party from 10 friends total.

Answer

$1, 10, 45, 120, \ldots$

Solution

The number of ways to invite 0 friend from a group of 10 friends is

$$\binom{10}{0} = 1.$$

The number of ways to invite 1 friend from a group of 10 friends is

$$\binom{10}{1} = 10.$$

If we repeat this process, we observe that the sequence is

$$1, 10, 45, 120, \ldots,$$

and is finite with length 11.

Problem 7.2 For each of the following, (i) come up with a way to describe the sequence and (ii) write out a few more terms. Note, your description does *not* need to be a formula.

(a) $3, 1, 4, 1, 5, \ldots$.

Answer

$9, 2, 6, 5, 3, \ldots$

Solution

The sequence consists of the digits in the decimal representation of π.

Therefore, the next few terms are:

$$9, 2, 6, 5, 3, \ldots.$$

(b) $1, 2, 9, 28, 65, \ldots$.

Answer

$126, 217, 344, \ldots$

Solution

The nth term of the sequence is $n^3 + 1$.

Therefore, the next few terms are:

$$126, 217, 344, \ldots.$$

(c) $3, 7, 15, 31, 63, \ldots.$

Answer

$127, 255, 511, \ldots$

Solution

Each term is twice the previous term plus 1.

Therefore the next few terms are:

$$127, 255, 511, \ldots.$$

Problem 7.3 Arithmetic sequences are given below, in one of three ways: (i) the first few terms of the sequence, (ii) the formula, or (iii) the recursive formula. Give the other 2 ways of describing the sequence. (That is, if the recursive formula is given, write out the first few terms and give the general formula for the sequence.)

(a) $a_0 = 5, a_{n+1} = 8 + a_n$.

Answer

$5, 13, 21, 29, \ldots; a_n = 5 + 8n$

Solution

$5, 13, 21, 29, \ldots; a_n = 5 + 8n$.

(b) $3, 5, 7, \ldots.$

Answer

$a_0 = 3, a_{n+1} = a_n + 2$ or $a_n = 3 + 2n$

Solution

$a_0 = 3, a_{n+1} = a_n + 2$ or $a_n = 3 + 2n$.

(c) $a_n = 6 - 5n$.

Answer

$6, 1, -4, -9, \ldots; a_0 = 6, a_{n+1} = a_n - 5$

Solution

$6, 1, -4, -9, \ldots; a_0 = 6, a_{n+1} = a_n - 5$.

Problem 7.4 Geometric sequences are given below, in one of three ways: (i) the first few terms of the sequence, (ii) the formula, or (iii) the recursive formula. Give the other 2 ways of describing the sequence. (That is, if the recursive formula is given, write out the first few terms and give the general formula for the sequence.)

(a) $2, 4, 8, \ldots$.

Answer

$a_0 = 2, a_{n+1} = 2 \cdot a_n; a_n = 2 \cdot 2^n$

Solution

$a_0 = 2, a_{n+1} = 2 \cdot a_n; a_n = 2 \cdot 2^n$.

(b) $a_0 = -3, a_{n+1} = -2 \cdot a_n$.

Answer

$-3, 6, -12, \ldots; a_n = -3 \cdot (-2)^n$

Solution

$-3, 6, -12, \ldots; a_n = -3 \cdot (-2)^n$.

(c) $a_n = 4^n$.

Answer

$1, 4, 16, \ldots; a_0 = 1, a_{n+1} = 4 \cdot a_n$

Solution

$1, 4, 16, \ldots; a_0 = 1, a_{n+1} = 4 \cdot a_n$.

Problem 7.5 More Complicated Recursive Sequences

(a) The *Fibonacci* sequence is a sequence $F_0, F_1, F_2, \ldots$ such that $F_0 = 0$, $F_1 = 1$, $F_{n+1} = F_n + F_{n-1}$. That is, the next term in the sequence is the sum of the previous two terms. Write out the first 11 terms of the sequence.

Answer

$0, 1, 1, 2, 3, 5, 8, 13, 21, 34, 55$

Solution

$0, 1, 1, 2, 3, 5, 8, 13, 21, 34, 55$.

(b) Suppose a sequence starts $G_0 = 2$, $G_1 = 1$, $G_{n+1} = 2 \times G_n - G_{n-1}$. That is, multiply the previous term by 2 and subtract the term before that. Calculate the first 8 terms of the sequence.

Answer

$2, 1, 0, -1, -2, -3, -4, -5$

Solution

Given that the first two terms of the sequence is 2 and 1, we see that the third term of the sequence is

$$2 \times 1 - 2 = 0.$$

Therefore, the fourth term of the sequence is

$$2 \times 0 - 1 = -1.$$

Repeating the above procedure yields that the next 5 terms of the sequence is,

$$0, -1, -2, -3, -4.$$

(c) Define a sequence generated by the following: start with 12 and divide by 2 if the number is even or take 3 times the number plus 1 if the number is odd. List out the first 10 terms of the sequence.

Answer

12, 6, 3, 10, 5, 16, 8, 4, 2, 1

Solution

Since the first term is even, the second term is

$$12 \div 2 = 6.$$

Since the second term is even, the third term is

$$6 \div 2 = 3.$$

Since the third term is odd, the fourth term is

$$3 \times 3 + 1 = 10.$$

Implementing the pattern, we get:

$$12, 6, 3, 10, 5, 16, 8, 4, 2, 1.$$

Problem 7.6 Answer the following.

(a) In an arithmetic sequence, if the fifth term of the sequence is 5 and the tenth term of the sequence is 15, what is the first term?

Answer

-3

Solution

The difference between the fifth term and tenth term is

$$15 - 5 = 10.$$

As these are 5 terms apart in the arithmetic sequence, we see the common difference must be

$$10 \div 5 = 2.$$

Hence the first term, which is 4 terms earlier than the fifth, is

$$5 \times 2 \times -4 = -3.$$

(b) In a geometric sequence, if the fifth term of the sequence is 4 and the seventh term of the sequence is 9, what is the tenth term?

Answer

$\frac{243}{8}$

Solution

The ratio between the fifth term and seventh term is

$$\frac{9}{4}.$$

Since these are 2 terms apart in the geometric sequence, we see the common ratio must be

$$\sqrt{\frac{9}{4}} = \frac{3}{2}.$$

Thus the fenth term, which is 3 terms after the seventh, is

$$9 \times \left(\frac{3}{2}\right)^3 = 9 \times \frac{27}{4} = \frac{243}{8}.$$

Problem 7.7 Calculate the Following Sums

(a) What is the sum of the first 100 positive integers?

Answer

5050

Solution

Consider adding the sequence forwards and backwards:

$$
\begin{array}{rcrcccrcr}
  & 1   & + & 2   & + & \cdots & + & 99 & + & 100 \\
+ & 100 & + & 99 & + & \cdots & + & 2   & + & 1   \\
\hline
= & 101 & + & 101 & + & \cdots & + & 101 & + & 101
\end{array}
$$

Since there are 100 terms in total, this sum is

$$101 \times 100 = 10100.$$

Remembering that this sum is actually twice what we want, our final sum is equal to

$$101 \times 100 \div 2 = 5050.$$

(b) Suppose you have 30 terms of the arithmetic sequence:

$$3, 8, 13, 18, \ldots, 148.$$

What is the sum of these 30 terms?

Answer

2265

Solution

Consider the trick we saw in the previous problem: Consider adding the sequence forwards and backwards:

$$
\begin{array}{rcrcccrcr}
  & 3   & + & 8   & + & \cdots & + & 143 & + & 148 \\
+ & 148 & + & 143 & + & \cdots & + & 8   & + & 3   \\
\hline
= & 151 & + & 151 & + & \cdots & + & 151 & + & 151
\end{array}
$$

Since this is twice the actual sum, we get

$$30 \times 151 \div 2 = 2265$$

as our answer.

Problem 7.8 Let a_n be the number of $n+1$ digit numbers made up of $0,2,4,6,8$.

(a) Write out a formula for a_n.

Answer

4×5^n

Solution

The first digit is either $2,4,6,8$, and ever other digit is $0,2,4,6,8$.

(b) Write out a recursive formula for a_n.

Answer

$a_0 = 4, a_{n+1} = 5 \times a_n$

Solution

Note this is a geometric sequence.

Problem 7.9 Let a_n (for $n \geq 1$) denote the number of ways to write n as the sum of 1's and 2's. For example, $3 = 1+1+1 = 2+1 = 1+2$ so $a_3 = 3$.

(a) Write out the first few terms of the sequence and guess a recursive formula for the sequence. (You do not need to prove your guess.)

Answer

$a_{n+1} = a_n + a_{n-1}$

Solution

The first few terms are $1,2,3,5,8$. Note this is roughly the Fibonacci sequence.

(b) Use your guess in (a) to calculate a_{10}.

Answer

89

Solution

89.

Problem 7.10 Suppose you have n friends. You want to invite some of them out to dinner. You will invite at least one friend, but not all of the friends.

(a) How many different ways can you invite a group of friends for $n = 1, 2, 3, 4$.

Answer

$0, 2, 6, 14$

Solution

See the general explanation in part (b).

(b) Find a general formula with n friends.

Answer

$2^n - 2$

Solution

Use complementary counting. With no restrictions there are 2^n ways of inviting a group of friends (each friend yes or no). We then subtract the two ways (no friends or all friends) we want to avoid.

(c) Let the answers above form a sequence $a_1, a_2, \dots$. Find a recursive formula for a_n.

Answer

$a_1 = 0, a_{n+1} = 2a_n + 2$

Solution

$a_1 = 0, a_{n+1} = 2a_n + 2$.

8   Solutions to Chapter 8 Examples

Problem 8.1 Suppose you roll 2 four-sided dice. Let A be the event that the first die is a 4, and B the event that the sum of the two rolls is 6.

(a) Write out a finite sample space Ω so that every outcome is equally likely. What is $n(\Omega)$?

Answer

$\Omega = \{(1,1),(1,2),(1,3),(1,4),(2,1),(2,2),\ldots,(4,4)\}; n(\Omega) = 16$

Solution

The list of all possible outcomes from rolling 2 four-sided dice is given below:

$$(1,1),(1,2),(1,3),(1,4),(2,1),(2,2),\ldots(4,4).$$

There are 16 total outcomes.

(b) Calculate $P(A)$.

Answer

$\dfrac{1}{4}$

Solution

We have

$$A = \{(4,1),(4,2),(4,3),(4,4)\}$$

so $n(A) = 4$. Hence

$$P(A) = \frac{n(A)}{n(\Omega)} = \frac{4}{16} = \frac{1}{4}.$$

(c) Calculate $P(B)$.

Answer

$\dfrac{3}{16}$

Solution

We have

$$B = \{(2,4),(3,3),(4,2)\} \Rightarrow n(B) = 3$$

so

$$P(B) = \frac{n(B)}{n(\Omega)} = \frac{3}{16}.$$

(d) Calculate $P(A \cap B)$.

Answer

$\dfrac{1}{16}$

Solution

Note that $A \cap B = \{(4,2)\}$ so $n(A \cap B) = 1$ and

$$P(A \cap B) = \frac{n(A \cap B)}{n(\Omega)} = \frac{1}{16}.$$

Problem 8.2 Suppose you flip a fair coin 6 times.

(a) Describe the sample space Ω. Find $n(\Omega)$.

Answer

64

Solution

Ω is the set of all words of length 6 using the letters H, T, so $n(\Omega) = 2^6$.

(b) Find the probability of exactly 5 heads.

Answer

$\dfrac{3}{32}$

Solution

If A is the event that we get exactly 5 heads,

$$n(A) = \binom{6}{5} = 6,$$

so

$$P(A) = \frac{\binom{6}{5}}{2^6} = \frac{6}{64} = \frac{3}{32}.$$

(c) Find the probability of at least one tails.

Answer

$$\frac{63}{64}$$

Solution

Note there is only 1 way to get 0 tails. If A is the event that we get at least one tails, then $n(A) = 64 - 1 = 63$ (using complementary counting).

(d) Find the probability of no two heads in a row and no two tails in a row.

Answer

$$\frac{1}{32}$$

Solution

Note the only possibilities are $HTHTHT, THTHTH$.

Problem 8.3 Suppose you randomly pick a point on the number line between 0 and 4.

(a) What is the probability the number is greater than 2?

Answer

$$\frac{1}{2}$$

Solution

The length of what we want is 2, and the total length is 4.

$$\frac{2}{4} = \frac{1}{2}$$

(b) What is the probability the number is greater than or equal to 2?

Answer

$\frac{1}{2}$

Solution

The length of what we want is still 2, and the total length is 4.

$$\frac{2}{4} = \frac{1}{2}$$

(c) What is the probability the number is equal to 2?

Answer

0

Solution

Note the point 2 has zero length!

Problem 8.4 Suppose Bill has a dart board with radius 4 feet. Whenever Bill throws a dart, it randomly lands somewhere on the board.

(a) What is the probability that Bill's dart lands within 2 feet of the center of the board?

Answer

$\frac{1}{4}$

Solution

The area of the inner circle is $\pi 2^2 = 4\pi$ and the full circle is $\pi 4^2 = 16\pi$.

$$\frac{\pi 2^2}{\pi 4^2} = \frac{1}{4}$$

(b) What is the probability Bill's dart lands between 2 and 3 feet from the center?

Answer

$$\frac{5}{16}$$

Solution

The region we want is inside a circle of radius 3, and outside a circle of radius 2, so has area $\pi 3^2 - \pi 2^2 = 5\pi$. We then divide this by the full circle.

$$\frac{\pi 3^2 - \pi 2^2}{\pi 4^2} = \frac{5}{16}$$

Problem 8.5 A dealer starts with only the 4 aces (one of each suit) from a deck of cards. They deal you 2 of the cards. Let A be the event that you get one heart and one diamond. Let B be the event that you get a spade.

(a) Assume the cards are dealt to you in order (that is, a first card and a second card). Find $P(A)$ and $P(B)$. Hint: You may want to write out a sample space.

Answer

$$P(A) = \frac{1}{6}, P(B) = \frac{1}{2}$$

Solution

The sample space Ω has size $4 \cdot 3 = 12$. $n(A) = 2$ (the 2 orders of the cards) and $n(B) = 2 \cdot 3$ (there are 3 choices for which other card to get, and then 2 orders).

$$P(A) = \frac{2}{12} = \frac{1}{6}, P(B) = \frac{6}{12} = \frac{1}{2}$$

(b) Assume the cards are not dealt in order (that is, you are dealt two cards at once). Find $P(A)$ and $P(B)$. Hint: You may want to write out a sample space.

Answer

$$P(A) = \frac{1}{6}, P(B) = \frac{1}{2}$$

Solution

Now the the sample space Ω has size $\binom{4}{2} = 6$. $n(A)$ is now 1, and $n(B) = 3$. (We no longer have orders to worry about.)

$$P(A) = \frac{1}{6}, P(B) = \frac{3}{6} = \frac{1}{2}$$

(c) Compare your answers in parts (a) and (b). Can you explain the outcome?

Solution

They are the same. When we are picking without replacement, as long as we are consistent about order or not order, we will get the same answer for probability.

Problem 8.6 Suppose $\Omega = \{1, 2, 3, 4, 5, 6\}$, and $P(1) = P(3) = .2, P(2) = P(4) = .1, P(5) = .15$.

(a) Find $P(6)$.

Answer

.25

Solution

$P(6) = 1 - .2 - .1 - .2 - .1 - .15 = .25$.

(b) Find the probability you get a prime number?

Answer

.45

Solution

The primes are $2, 3, 5$ so the probability of getting a prime is $.1 + .2 + .15$.

(c) Let $A = \{1, 3, 5\}, B = \{2, 3\}$. Verify that $P(A \cup B) = P(A) + P(B) - P(A \cap B)$. (That is calculate each term separately and check the formula.)

Solution

$P(A \cup B) = .65, P(A) = .55, P(B) = .3, P(A \cap B) = .2$. It is true that $.65 = 55 + .3 - .2$.

Problem 8.7 Suppose you have 5 red and 8 green balls. You pick 5 without replacing the balls. For each of the events below, find the probability. Note: It is best to think of all 5 balls being picked at once.

(a) What is the probability you get 3 green and 2 red balls?

Answer

$\frac{560}{1287}$

Solution

The total number of outcomes in the sample space is

$$\binom{13}{5} = 1287$$

since we are interested in choosing 5 balls from a group of 13 balls. Of the 5 balls, there are

$$\binom{5}{2} = 10$$

ways to decide which red balls to get and

$$\binom{8}{3} = 56$$

ways to decide which green balls to get. Therefore, the probability is

$$\frac{\binom{5}{2}\binom{8}{3}}{\binom{13}{5}} = \frac{56 \times 10}{1287} = \frac{560}{1287}$$

(b) What is the probability that all the balls you pick are the same color?

Answer

$\frac{19}{429}$

Solution

The total number of outcomes in the sample space is

$$\binom{13}{5} = 1287$$

since we are interested in choosing 5 balls from a group of 13 balls. We have two cases to consider.

The first case is when all 5 balls are red. There are

$$\binom{5}{5} = 1$$

way of choosing 5 balls to be red.

The second case is when all 5 balls are green. There are

$$\binom{8}{5} = 56$$

way of choosing 5 balls to be green. Therefore, the desired probability is

$$\frac{\binom{5}{5} + \binom{8}{5}}{\binom{13}{5}} = \frac{57}{1287} = \frac{19}{429}.$$

Problem 8.8 A, B are events. Answer the following. Drawing a Venn Diagram may help!

(a) Suppose $P(A) = .6, P(B) = .7$. What are the maximum and minimum possible values of $P(A \cap B)$?

Answer

.6, .3

Solution

If A is contained in B, then $P(A \cap B) = P(A) = .6$ (and this is the maximum). The minimum overlap will occur when $P(A \cup B) = 1$. In this case $P(A \cap B) = .6 + .7 - 1 = .3$.

(b) Suppose $P(A) = .5, P(B) = .6, P(A^c \cap B^c) = .2$. Find $P(A \cap B^c)$ and $P(B \cap A^c)$.

Answer

$.2, .3$

Solution

$1 = P(A^c \cap B^c) + P(A) + P(B \cap A^c)$ so $P(B \cap A^c) = .3$. From here it follows that $P(A \cap B) = .3$ and thus $P(A \cap B^c) = .2$.

Problem 8.9 Suppose Jack and Jill decide to meet up for dinner at 5:30 PM. Unfortunately, with traffic, they each arrive (separately) sometime at random between 5 and 6 PM. Hint: think of graphing when they arrive on a square. What is the probability they both arrive on time? (Be careful what this means!)

Answer

$\dfrac{1}{4}$

Solution

If we view the sample space as a 1×1 square, it has area 1. If they both arrive on time (both before 5:30 PM, or in the first half of the hour), we get a $\dfrac{1}{2} \times \dfrac{1}{2}$ square in the lower-left of the bigger square. Therefore, the area of what we want is $\dfrac{1}{4}$.

$$\frac{1/4}{1} = \frac{1}{4}$$

Problem 8.10 Suppose you have two sticks, one of length 1 and the other of length 2. You now get a third stick whose length is randomly chosen between 0 and 5. What is the probability you can make a triangle with the three sticks?

Answer

$\dfrac{2}{5}$

Solution

Let x be the length of the third stick. By the triangle inequality we need $1+2>x$, $1+x>2$, and $2+x>1$. Hence we need $1<x<3$. This interval has length 2, out of a total of 5.

9 Solutions to Chapter 9 Examples

Problem 9.1 Given the list of numbers $1, 4, 6, 4, 3, 8, 2$ find the

(a) mean.

Answer

4

Solution

We have $1 + 4 + 6 + 4 + 3 + 8 + 2 = 28$, and $28/7 = 4$.

(b) median.

Answer

4

Solution

Sorted the list is $1, 2, 3, 4, 4, 6, 8$.

(c) mode.

Answer

4

Solution

4 is the only number that appears multiple times.

Problem 9.2 Consider the list of numbers $1, 4, 6, 4, 3, 8, 2$.

(a) What (single) number do you need to insert to the list so that the mean changes to 5?

Answer

12

Solution

For a mean of 5, we need the 8 numbers to sum to 40. Hence the new number must be 12.

(b) Suppose instead that a new number is inserted. If the new number is an integer, what are all the possibilities for the median?

Answer

3.5, 4

Solution

If the number added is greater than or equal to 4, the median is $(4+4)/2 = 4$. Otherwise the median is $(3+4)/2 = 3.5$.

Problem 9.3 Suppose you have a list of 5 integers each between 1 and 10 (inclusive).

(a) How many different lists are possible? Answer this with and without order.

Answer

$$10^5; \binom{5+10-1}{5}$$

Solution

With order it is just 10 choices for each element. Without order we use stars and bars.

(b) How many different means are possible?

Answer

46

Solution

Note that counting means is the same as counting sums. The smallest sum is 5 and the largest sum is 50. Every sum in between is possible, so there are $50 - 5 + 1$ different sums and so $50 - 5 + 1$ different means.

(c) Suppose the list is picked in order (first, second, etc.) one element at a time at random. What is the probability that the mean of the list is 2.

Answer

$$\binom{9}{5}/10^5 = \frac{126}{10^5} = \frac{63}{50000}.$$

Solution

In total there are 10^5 outcomes. If the mean of the list is 2, the numbers must sum to 10. There are $\binom{9}{5}$ non-negative solutions to $a+b+c+d+e = 10$, which each correspond to a different list.

Problem 9.4 The SAT is a renowned high school standardized test administered for college admissions purposes. The scores are in multiples of 10 (maximum 800 per section) and the total score is determined by adding the critical reading subscore and the math subscore. Of the 50 students that scored above 600 on either the critical reading or math section of the SAT, 35 students scored above 600 on the math section and 25 students scored above 600 on the critical reading section. What is the maximum average total score of all 50 students?

Answer

1440

Solution

Given that there are 50 students that scored above 600 on either the critical reading or math section of the SAT, the overlap of students that score above 600 on either sections represent the number of students that score above 600 on both sections.

Therefore, there are

$$(35+25) - 50 = 10$$

students that score above 600 on both critical reading and math sections of the SAT.

The maximum total score that these 10 students can score is

$$800 + 800 = 1600,$$

and the maximum total score that the remaining 40 students can score is

$$800 + 600 = 1400.$$

Therefore, the sum of the 50 SAT total scores is

$$40 \times 1400 + 10 \times 1600 = 72000$$

and the average of the 50 SAT total scores is

$$72000 \div 50 = 1440.$$

Problem 9.5 Given below is a line graph indicating the price of a stock share in the market:

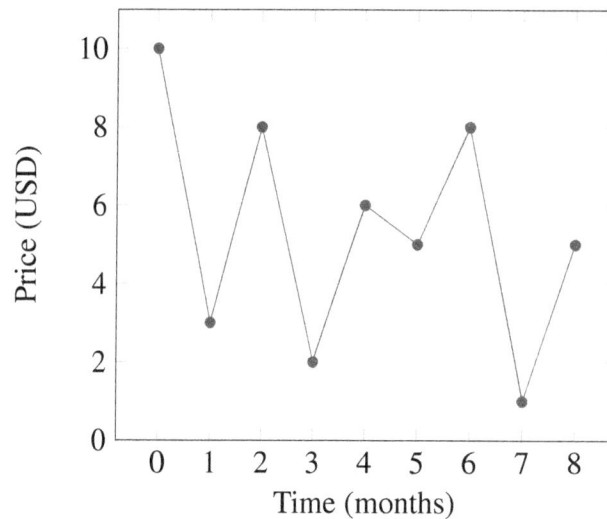

What is the average change of price of the stock every month?

Answer

-0.675

Reading the above line graph, we note that the change of the stock price from month 0 to month 1 is

$$3 - 10 = -7.$$

The change of the stock price from month 1 to month 2 is

$$8 - 3 = 5.$$

Generating all of the changes of the price yields the following list of numbers:

$$-7, 5, -6, 4, -1, 3, -7, 4.$$

Therefore, the average of the list of numbers above is:

$$\frac{-7+5-6+4-1+3-7+4}{8} = -0.675$$

Problem 9.6 Given below is a bar graph representing the ages of all students in a class.

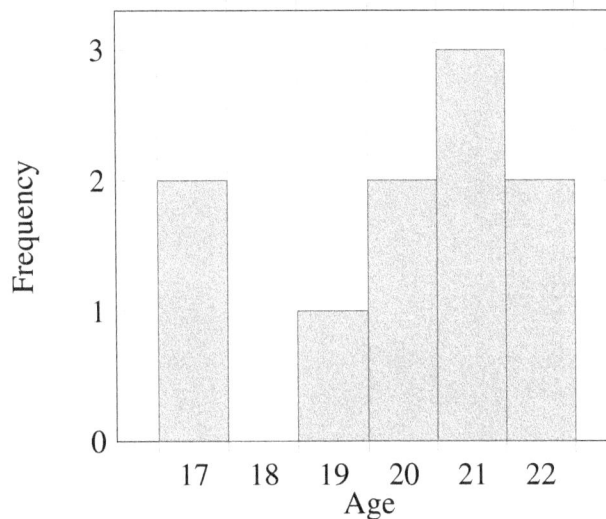

Find the median age of the students in the class.

Answer

20.5

Solution

Reading the above bar graph, we note that there are 2 students of age 17, 1 student of age 19, 2 students of age 20, 3 students of age 21, and 2 students of age 22.

Listing out all of the ages of the students, we have the following list:

$$17, 17, 19, 20, 20, 21, 21, 21, 22, 22$$

Starting from the ends of the list, we cross each number off one by one until we are left with the two middle numbers 20 and 21. The average between 20 and 21 represents the median age of the students in the class. This average is computed as follows:

$$\frac{20 + 21}{2} = 20.5$$

Problem 9.7 Suppose you have the list of integers $0, 0, 2, 2, 2, 6, 8, 8, 9, 10$. You randomly pick one element in the list.

(a) What is the mode? What is the probability of picking the mode?

Answer

$P(2) = .3$

Solution

The mode is 2, the the probability of picking it is $3/10$. Note that this is the number most likely to be picked.

(b) What is the median of the list? What is the probability of picking the median?

Answer

$P(4) = 0$

Solution

The median is $(2+6)/2 = 4$. However, since 4 is not in the list, it has probability 0 of being picked.

(c) What is the probability of picking a number larger than the median?

Answer

0.5

Solution

Note half the numbers are above the median.

Problem 9.8 Pretend that everyone likes all the Harry Potter books equally, so every person has an equal chance of picking any of the books as their favorite. Suppose you poll 5 people.

(a) What is the probability that the list of favorite books is $5, 2, 7, 1, 6$?

Answer

$\dfrac{1}{7^5}$

Solution

There are 7^5 total outcomes, and only one listed above.

(b) What is the probability that the median of the favorite books (of the 5 people polled) is the last book? Hint: This means 3, 4, or all 5 chose the last book as their favorite.

Answer

$$\frac{\binom{5}{3} \cdot 6^2 + \binom{5}{4} \cdot 6 + \binom{5}{5} \cdot 1}{7^5}$$

Solution

Note that in each case we need to decide who chose the last book as their favorite, and which of the other 6 books the others chose as their favorite.

Problem 9.9 Provided below are the heights (in inches) of 5 randomly selected adult males in the U.S.:

$$70.2, 73.9, 76.9, 69.4, 72.2$$

Provided below are the heights (in inches) of 5 randomly selected adult females in the U.S.:

$$70.6, 73.0, 65.0, 68.0, 61.9$$

The following questions regarding the heights of U.S. adult males and females are based on the data collected above.

(a) Find the mean height of U.S. adult males.

Answer

72.52

Solution

For males, the mean is calculated as follows:

$$\frac{70.2 + 73.9 + 76.9 + 69.4 + 72.2}{5} = 72.52$$

(b) Find the mean height of U.S. adult females.

Answer

67.7

Solution

For females, the mean is calculated as follows:

$$\frac{70.6 + 73.0 + 65.0 + 68.0 + 61.9}{5} = 67.7$$

(c) Find the median height of U.S. adult males.

Answer

72.2

Solution

Ordering the heights, we have the following:

$$69.4, 70.2, 72.2, 73.9, 76.9$$

Therefore, the median of the list is 72.2.

(d) Find the median height of U.S. adult females.

Answer

68.0

Solution

Ordering the heights, we have the following:

$$61.9, 65.0, 68.0, 70.6, 73.0$$

Therefore, the median of the list is 68.0.

(e) Is it true that it is more likely for a randomly selected U.S. adult male to be taller than a randomly selected U.S. adult female? Why?

Answer

True

Solution

Since the mean height of U.S adult males is greater than the mean height of U.S. adult females, we can conclude that it is more likely for a randomly selected U.S. male to be taller than a randomly selected U.S. female.

Problem 9.10 Suppose the following list represents the grades of a test in a math class:

$$60, 64, 78, 80, 90, 92, 94, 95, 97, 99$$

(a) Find the median score of the class.

Answer

91

Solution

Since there is an even amount of test scores, observe that two scores 90 and 92 are test scores that lie near the median score of the class. Therefore, the average of the two scores would represent the median score of the overall class.

Thus, the median score of the class is:

$$\frac{90+92}{2} = 91.$$

(b) Find the mean score of the class.

Answer

85

Solution

The mean score of the class is determined by finding the sum of all of the grades and dividing it by the number of grades. The sum is

$$60+64+78+80+90+92+94+95+98+99 = 850.$$

Therefore, the average is

$$850 \div 10 = 85.$$

(c) Suppose another 5 students in the class score 90 on the quiz. How does this effect the mean and median?

Answer

The median will decrease and the mean will increase.

Solution

If at least two scores of 90 are added to the list, then the median will decrease from 91 to 90.

The mean of the scores will increase. Specifically, the new sum will be,

$$850+5 \times 90 = 1300.$$

and the new mean will be,

$$1300 \div 15 \approx 86.67.$$

(d) Do you think the median or mean does a better job at describing the given data?

Answer

Median

Solution

Since more than half of the students got 90 or above, the median probably does a better job describing the data.